CAMBRIDGE COMPARATIVE PHYSIOLOGY

GENERAL EDITORS:

J. BARCROFT, C.B.E., M.A.
Fellow of King's College and Professor of
Physiology in the University of Cambridge
and
J. T. SAUNDERS, M.A.
Fellow of Christ's College and Lecturer in
Zoology in the University of Cambridge

THE COMPARATIVE PHYSIOLOGY
OF MUSCULAR TISSUE

T0296086

THE COMPARATIVE PHYSIOLOGY
OF MUSCULAR TISSUE

BY

A. D. RITCHIE, M.A.

Lecturer in Chemical Physiology in the Victoria
University of Manchester, formerly
Fellow of Trinity College
Cambridge

CAMBRIDGE
AT THE UNIVERSITY PRESS
1928

CAMBRIDGE
UNIVERSITY PRESS

University Printing House, Cambridge CB2 8BS, United Kingdom

Cambridge University Press is part of the University of Cambridge.

It furthers the University's mission by disseminating knowledge in the pursuit of education, learning and research at the highest international levels of excellence.

www.cambridge.org
Information on this title: www.cambridge.org/9781107502338

First published 1928
First paperback edition 2015

A catalogue record for this publication is available from the British Library

ISBN 978-1-107-50233-8 Paperback

CONTENTS

INTRODUCTION

The physiology of muscle may be studied in many different ways. In the first place we may deal with the muscular mechanism of the organism as a whole, with what may be called the *physiology of animal movement*, the study of which was begun by Borelli and has been pursued more recently by Marey and a few others. This subject has been treated from the comparative point of view by du Bois Reymond in Winterstein's *Handbuch der Vergleichenden Physiologie*. It is concerned with the variety of methods by which animals succeed in moving about. Allied to this subject are two questions; that of the static maintenance of form and posture, and the application of the *principle of dynamical similarity* to the statics and dynamics of the animal body; namely, the mechanical principles limiting the types of structure and locomotion, and the size, speed and endurance of animals of similar form. These questions are dealt with by D'Arcy Thompson in *Growth and Form*.

A second mode of approach still deals with the animal as a whole, not externally as a problem in mechanics, but as a *neuro-muscular system*, as an organism which reacts in characteristic ways to external changes. This, the study of animal *behaviour*, which may reasonably be said to have originated with the work of Lloyd Morgan, has since been pursued with vigour, particularly in America by zoologists and psychologists. Within this sphere of investigation comes the study of the behaviour of Protozoa and hence part at least of the study of amoeboid and ciliary processes.

We may next proceed to analyse the muscular response of the organism, treating it as a system built up of component mechanisms. The components will be the *reflexes*. Our knowledge of the reflex mechanism, it need hardly be mentioned, is chiefly the result of the labours of three physiologists and their schools—Pavlov, Magnus, and Sherrington. As far as the skeletal system of the higher animals is concerned, the *organ* or functional unit, out of which the whole organism is built up as a system, is the complex structure needed for a complete *reflex arc*. The muscle itself is only the *effector* half of the organ. The simplest organ is the functional unit needed for what Sherrington calls a "myotatic reflex", and consists of *muscle spindle, nervous connections* (at least two neurones) and a group of *muscle fibres*. On the other hand the heart is an *organ*, a complete functional unit. It is a self-exciting mechanism and needs the addition of no other structure to enable it to carry out its functions. The distinction must be borne in mind when the performance of the isolated heart is compared with that of the isolated skeletal muscle; the one is a whole, the other a fragment. This does not imply that the study of the isolated skeletal muscle is useless or misleading. On the contrary, it is a great technical advantage to be able to sever the exciting mechanism from what is excited because then excitation is under control. It is much harder to produce controlled excitation in the heart where there is already a self-contained exciting mechanism, as there is also in the muscular mechanism of the digestive organs. It so happens that the heart is an organ with a constant function and gives a constant response under suitable conditions. The stomach or intestine not only contains different types of muscle, but is liable to show a variable spontaneous activity.

A. J. Clark, in his book in this series, has treated the heart as an *organ* in the sense used above; that is to say he has discussed its working as a functional unit and as part of the economy of the whole body. The physiology of the various muscular organs containing unstriped muscle could be treated in the same way, though there is hardly enough information for comparative treatment to be possible. As has been mentioned already, the physiology of skeletal muscles as organs resolves itself into the study of reflex action.

The aspects of muscle physiology enumerated above are not dealt with in this book, which is concerned with the last remaining aspect. The title, *Comparative Physiology of Muscular Tissue*, is intended to convey this limitation. The aim may be said to be to describe the characters and functions of muscle cells. A muscle or group of muscles is considered simply as an aggregate of muscle fibres, not as a complex system forming an organ of the body. The limitations of this method are obvious. The muscle is not merely an aggregate of muscle fibres, and the fibres are not uniform, so that the results obtained represent a resultant of the working of all structures present in the tissue. It is true that a beginning has been made towards a study of the individual muscle fibre, but there is not much information from this source as yet. For most of our information we have to rely upon investigation of whole muscles or groups of muscles. The data may be crude, but are sufficient for a first approximation.

If we treat the animal body analytically, it is clear that there must be ultimate functional units, physiological atoms. These may be taken to be the muscle fibres in the case of muscle, since even if the fibres are capable of further subdivision there is no method at present known of investigating smaller units, except for rather dubious histological methods. Therefore our knowledge of the muscular system as a whole should begin with the knowledge of the smallest parts, the fibres, and the simplest aggregates of them, the anatomically separable muscles. It is this information with which the book is concerned. Some physiologists have insisted that the whole organism is not a simple aggregate of its parts. This is true enough, but it sometimes leads to a belief that therefore the investigation of the parts is of no importance. The whole organism is presumably some function (in the mathematical sense) of its parts though not a simple function. It is necessary therefore to study the organism as a whole separately to determine the laws of the whole, as well as those of the parts. But to refuse to take account of the laws of the parts is merely to neglect relevant and valuable information. The importance of the relation is seen if we consider that the information obtained from a study of the respiratory exchange in man during muscular exercise, that is the metabolism of the whole organism, was relatively trivial until it could be correlated with the knowledge of the processes in the muscle fibre which had been obtained by experiments on isolated frog's muscle.

Let us assume that the physiology of the cell is the basis of all physiology, even though the whole organism is not a simple sum of its parts and it is not always clear how the parts are related to the whole. Of the different types of cell that compose the bodies of higher animals the muscle cells are important quantitatively in that they constitute more than half the active living material. They are also important qualitatively in that the higher organisms are essentially neuro-muscular systems. The animal *is* its sense organs, nervous system and skeletal muscles; the rest of the body consists of auxiliary organs and passive mechanical framework. Even of the auxiliary organs part is muscular. Moreover it so happens that there is more information about the processes in the muscle cells than there is about processes in any others. The reason for this is simply that no other tissue can be made to produce so many measurable effects which can be correlated with one another.

This one-sided development of cell physiology is unfortunate in that it is often difficult to say whether some process found going on in a muscle is a specific muscular process or something common to all excitable cells.

In the following pages a certain thesis is developed, and it may be as well to confess at once what this is. The investigation of isolated frog's muscle has revealed, in rough outline at least, certain physical and chemical changes in the muscle fibre that are the causes of what appears as the *simple muscle twitch*. The results can be applied to a considerable range of types of muscle, possibly to all muscles. *On the present evidence there is no need to postulate any mechanism other than that underlying the simple muscle twitch*. Different muscles differ profoundly in the speed of the processes but not apparently in the kind of process. The phenomena of *tonus* and of so-called *catch muscles* do not need the assumption of a special mechanism. As regards excitation, however, there are two distinct types of muscle. The one kind is self-exciting (e.g. vertebrate heart), the other is not (e.g. vertebrate skeletal muscle).

Although most of the oxidation in the body takes place in the muscles, nothing will be said here as to the chemical mechanisms of animal oxidations. There are two reasons for this. First, although there is now much information as to various parts of the oxidative mechanism, there is little or none as to how these parts fit in to make up the whole oxidizing system. The information is fragmentary. Secondly, there is no reason to believe that there is anything peculiar to muscle, as distinct from other tissues, in the oxidations that occur, except perhaps for the predominant part played by carbohydrate metabolism.

The general arrangement of the book and the subjects dealt with are as follows. In Chapter I the general characters of some of the best-known types of muscular tissue are described along with some indications as to the distinguishing features by which different types of muscle can be classified. The chemical composition of muscles is then dealt with, and finally some chemical changes that occur in muscle and that are not necessarily associated with muscular activity. Chapter II is concerned with the chemical changes accompanying and underlying muscular activity. The processes that have been found to occur in isolated frog's muscle are first discussed and then processes in other types of muscle. There is no evidence as yet that the processes in any muscle differ qualitatively from those in frog's skeletal muscle, the "type" muscle. But there are considerable quantitative differences, the significance of which is discussed. In Chapter III the physical processes underlying muscular activity are dealt with. Following this there is a short discussion of certain theories of muscular contraction. The third section of the chapter is concerned with the subject of muscular tone, to determine whether there is any reason to suppose it is the result of some special process in the muscle differing fundamentally from the contractions that can be produced experimentally in isolated tissues. Lastly, in Chapter IV there is a very brief account of the action of electrolytes and other substances in the environment of the muscle cell.

Chapter I

GENERAL CHARACTERS OF MUSCULAR TISSUE

§ I. *TYPES OF MUSCLE*

A muscle consists essentially of a bundle of elongated fibres, each of which may be considered as a syncytium of cells and each of which is functionally a single unit. Generally each fibre is functionally distinct from its neighbours so that any one fibre can be excited by itself without the excitation spreading beyond that fibre. The characteristic sign of the special function of muscular tissue consists in shortening and developing tension on excitation by an appropriate stimulus and in subsequent relaxation to the original length and tension. Other functions may be superimposed upon this kind of response, but it may be treated as the type.

Different muscles differ *morphologically* with respect to:

(1) Their origin in the embryo;

(2) Their anatomical position and structure;

(3) Their histological appearance;

(4) Their chemical composition;

(5) Their nerve supply;

functionally with respect to:

(6) The type of stimuli to which they are sensitive;

(7) The type of response to stimulation.

(6) and (7) might be further subdivided. Thus muscles may differ in their time scale, as for example insect's wing muscles and tortoise's heart muscle. More important still, they may differ according to whether they are spontaneously active, that is, whether internal changes may act as stimuli, or not, as for example frog's heart muscle and skeletal muscle.

In attempting to classify muscles it is convenient to consider first vertebrate muscle, particularly mammalian muscle, about which we have considerable knowledge, and consider two strongly contrasted types; then it will be found that most types of muscular tissue in the animal kingdom resemble one or other type or are intermediate. First then consider two types of mammalian muscle which we may call *skeletal* and *visceral*. These are distinct embryologically and anatomically, if we confine our attention to typical cases and neglect doubtful or intermediate ones. They are also distinct according to the other criteria enumerated.

SKELETAL MUSCLES OF HIGHER VERTEBRATES

Taking *skeletal* muscle first, as it is found in the body. A Mammal or any other *walking* animal is mechanically an articulated girder structure, of which the skeleton forms the system of *compression members* and the muscles (along with certain other structures) the *tension members*. The muscles are not only the *prime movers* of the body but also the *tension members* on which the body as a static structure depends. That is to say, they are necessary to maintain bodily posture in the gravitational field. The muscular tensions, *tonus*, by which posture is maintained, are smaller than the tensions developed during movement, but they are maintained for long periods with small expenditure of energy. In considering skeletal muscle, and any other sort of muscle, the function of maintaining *tone* cannot be neglected. Mammalian skeletal muscles differ in type among themselves, are not homogeneous structures (excluding non-muscular material such as tendon and blood vessels), and lastly are part of a highly specialized neuro-muscular system. Taking these points in order, muscles have been distinguished as red and white, and the distinction is found to correspond more or less to the amount of use the muscle has, as for example the wing muscles of fowl and pigeon or the leg muscles of hare and rabbit, and also to certain histological differences. The differences are not constant; for instance, wild rabbit's muscle is redder than tame and it is said that among men, the muscles of athletes are redder than those of old or sedentary individuals. Further, histologically all mammalian muscles are seen to be mixed, and the increase that follows use is seen to be due to hypertrophy of existing cells not to growth of new ones. Continual use apparently tends to bring about an increase in the "sarcoplasm" of the muscle fibre and among other things an increase of haemoglobin, the red pigment. Though one may still ask, Why are the frog's leg muscles white and the tortoise's red?

COMPLEX STRUCTURE OF SKELETAL MUSCLES

This brings us to another point, that histologically mammalian skeletal muscle is complex, as is nearly all muscle of similar character. Fibres, usually thin, with a granular appearance, can be distinguished from large, more translucent ones. The former are usually more pigmented and are more numerous in typical red muscle. There is a certain degree of correlation between pigmentation, granular fibres and constant use. Generally also the red type is more slowly moving (see Knoll, 1891; D. M. Needham, 1926). It must be pointed out however that Greene (1926) finds that the dark (red) muscles under the skin of the Pacific king salmon (*Onchorynchus tschawytschia*) are functionally degenerate compared with the pale (pink) muscles, but are full of fat and apparently act as fat depôts. Probably the dark muscle of other fishes is similar. Altogether too much stress should not be laid upon the distinction between red and white muscle or the two types of muscle fibre, except to recognize the fact that the muscle is a complex tissue.

In addition to the two types of fibre already mentioned there is a third most important structure belonging to the muscle tissue proper, the *muscle spindle*. This, as Sherrington has shown, is a sense organ developed in connection with a specialized muscle cell (see Cobb, 1925). The special significance of these structures in relation to muscle *tone* will be referred to later, suffice to say at present that it is characteristic of mammalian skeletal muscle to have no spontaneous activity apart from stimulation through its motor nerves. The reflex response of the muscle to passive tension is regulated by means of the muscle spindles as *affector organs*. Whereas visceral muscle represents an unspecialized, possibly primitive, system which is to a large extent self-regulating and only casually dependent on the central nervous system, the skeletal muscle is entirely bound up with it. This of course explains the fact that the isolated skeletal muscle only reproduces some of the activities of the intact organ; by isolating it we have destroyed the *afferent* side of the system of sense organ, nerve and muscle which exists in the body.

A further point to be mentioned is the still-vexed question of the *autonomic nerve supply* to skeletal muscle, on which subject see Hines (1927) and Fulton (1926, Chap. XVI). The nerve supply to a muscle carries (*a*) *myelinated fibres* which are *efferent* and come from the *anterior nerve root*, (*b*) *myelinated fibres* which are *afferent* from the *spindles* and go through the *posterior root* into the cord. In addition there are (*c*) *non-myelinated fibres*, not only those going to blood vessels but almost certainly to the muscle fibres themselves and ending in special types of nerve ending. Their function is still a matter of controversy, some workers suggesting that they are responsible for the maintenance of *tone* (see Cobb, 1925).

HISTOLOGY OF SKELETAL MUSCLES

A histological feature of skeletal, and some other muscles, to which much attention has been drawn in the past, is the appearance of *cross striation* in the fibres. The fact that many muscles are not cross striated shows that the structures giving this appearance, whatever they are, are not a necessary part of the contractile mechanism. They are more probably a special modification of the contractile mechanism for special functions. We may therefore dismiss all theories of muscular contraction based, as many have been, on the detailed microscopic appearance of the cross striation of certain kinds of muscle. It must be remembered too that the details of the cross striated structure are of a size near the limits of microscopic vision so that the visible appearances may be delusive. Haycraft (1891) was able to obtain all the appearances of cross striated structure, except those depending on the birefringence of the muscle, by taking casts in collodion from the surface of muscles. He considered that the various structures described by histologists were the optical results of a very fine surface varicosity of the fibres. These observations have been ignored by most histologists, but they seem to merit serious consideration.

In general, cross striation is most conspicuous in the most rapidly moving muscles, such as insect's wing muscles. The most plausible suggestion yet made is therefore that of Hill (1926) that cross striated muscle has an arrangement comparable to "baffle plates" set across the length of the fibre to reduce the amount of flow of the semi-liquid contents. Loss of energy by internal friction in the muscle is the chief source of lowered mechanical efficiency when the rate of movement is increased. If the muscle fibre were like a hollow tube with the walls thickened at intervals, this would account for the cross striated appearance and be compatible with Haycraft's observations, supposing there was a swelling corresponding to the "baffle plates". For further information as to the histology of muscle the reader should consult Schafer (1912).

FUNCTIONAL CHARACTER OF SKELETAL MUSCLES

Functionally it is characteristic of skeletal muscles in contrast to other types to show no spontaneous activity when denervated. They are not very extensible, and stretching does not act as a stimulus. They are stimulated by electric currents of short duration but require fairly high potentials to excite them; in particular they are more readily excited by single shocks from an induction coil than any other type of muscle. They have a short *refractory period* so that repeated stimulation of suitable frequency gives rise to fused contractions (*tetanus*). They can develop high tensions and their response to stimulation is rapid. In the last respect muscles from different animals and even different muscles from the same animal vary greatly, for example, the rapid twitch of frog's leg muscles and the much slower response of the muscles of the abdominal wall.

VISCERAL MUSCLES OF HIGHER VERTEBRATES

The visceral muscles are contrasted with the skeletal muscles in nearly every respect. The muscular tissue of the alimentary canal may be taken as the type. This consists of sheets or layers of contractile cells forming part of the wall of a tube or sac. The muscle cells in different layers may run longitudinally, circularly or diagonally. The general effect of their contraction is to exert pressure on the contents of the organ and change its shape. The muscles of this type are distinct from the skeletal muscles also in their origin in the embryo and their histological character; they are not cross striated. They have generally a double nerve supply which consists of *non-medullated* fibres of the *autonomic system*. They are nearly always spontaneously active apart from nervous stimulation, with a tendency to rhythmical movements and also to tonic contraction; some muscles produce one type of effect, rhythmic or tonic, more markedly than the other. Spontaneous activity and double innervation must be closely connected. Skeletal muscle which is inert by itself needs only excitation and therefore one set of nerve fibres. Spontaneously active muscle if it is to be controlled needs both inhibition and augmentation, therefore, two sets of fibres.

It is not necessary to decide whether the spontaneous activity of the musculature of the alimentary canal is really a function of the muscle fibres themselves or, as is possible, of the nerve net found in close contact with the layers of muscular tissue. The important point is that the organs and the experimentally separable parts of the organs are spontaneously active apart from external stimulation, nervous or otherwise. The minor question as to the exact *locus* of the self-excitatory process cannot be settled except by the actual separation of muscular and nervous elements.

A peculiarity of some visceral muscles is that conditions which excite when relaxed (passage of gas bubbles, mechanical stroking of surface, some kinds of electrical stimuli, adrenaline) produce relaxation when the muscle is in a state of tonic contraction (Evans, 1926, 2). This may be a consequence of double innervation, depending upon the inhibitor nerve endings being more excitable during prolonged activity and the augmentor nerve endings during rest. Compared with skeletal muscle the visceral muscular tissue is usually more extensible, and stretching acts as a stimulus, or at least increases excitability. Electric currents need to be of longer duration to stimulate, the response is generally more prolonged, and the tensions developed are much smaller.

So far we have been considering typical skeletal and visceral muscles of a Vertebrate, more particularly a Mammal, and emphasizing the differences; but there are a good many muscles in the body of an intermediate type. As Cobb (1926) says, "In fact it seems that no sharp lines can be drawn between the different types of muscle. *Smooth muscle* and *striated muscle* seem different enough but when one considers the muscles of the eyeball, larynx, tongue, diaphragm and heart it is difficult to see where to draw the dividing line between two obviously different extremes". And he goes on to quote some other authorities for his views. The view which will be maintained in what follows is that there is no *fundamental* difference between any muscles, but that the differences are differences of degree. The muscles of vertebrate limbs are specialized for rapid movement, the production of large tensions, complete central control and considerable delicacy of control, as witness the control of the human fingers. The muscles of the intestine are specialized or perhaps they are primitive, but at any rate they carry out slow movements involving small stresses, they are automatic for the most part and only controlled in a general way by the central nervous system. That is to say, the central nervous system augments or inhibits but the actual type and course of movement is dictated by local conditions and largely by the state of the tissue itself. Therefore we may say that muscles differ chiefly in respect to:

Time scale—rapidity of response as determined by time relations for electrical stimulation, latent period, refractory period, duration of the two phases of the mechanical response.

Intensity of response—measured by tension developed per unit cross section with maximal stimulation, or amount of chemical change per unit mass.

Degree of automaticity—including not only tendency to spontaneous activity but the kind of connection with the central nervous system.

Histological and other structural characters may be considered as of minor importance unless they are diagnostic of functional character. It must be remembered that muscles of widely different functional character are grouped together as "smooth muscle" for no better reason than mere absence of visible cross striation. The vertebrate heart is an interesting and highly specialized type of muscle intermediate in character between the skeletal and typical visceral muscle. An important feature of heart muscle is that the whole organ is functionally a *syncytium*, so that the excitation of one part tends to excite the whole organ. At the same time different regions contain muscle cells of very different physiological character as regards excitability and type of response— compare the spontaneously active tissue of the *sinu-auricular node* of the mammalian heart with the *ventricular* tissue.

The *retractor penis* of the dog is an unstriated muscle of an interesting type approaching that of skeletal muscle (Winton, 1927).

MUSCLES OF INVERTEBRATES

Leaving the vertebrate group of animals we find that the musculature of the Arthropods presents no great difficulty in classification, their limbs are moved by typical skeletal muscles and the intestines by typical visceral muscles, and there are intermediate types. The heart muscle in particular shows interesting variations. The heart of *Limulus* shows in its loss of automaticity an approximation to the skeletal type, while the crustacean heart, like that of the Vertebrate, approximates more to the visceral type. Outside these two great Phyla we find classification more difficult. The body muscles of the Cephalopods among the Molluscs and the adductors of the shell of the Lamellibranchs may be considered as skeletal and the rest visceral. When we consider the Annelid Worms there is little to go upon. The earthworm for instance possesses no skeleton and none of its muscle is striated. We might either (1) on general morphological grounds treat the body muscles as skeletal and the gut muscles as visceral, or (2) on histological grounds treat them all as visceral, or (3) consider that they all represent more primitive undifferentiated types. Without more complete investigations a choice cannot be made between these alternatives.

§ II. CHEMISTRY OF MUSCULAR TISSUE

The chemical composition of muscles is fairly characteristic, and in certain cases at least functional differences between different muscles can be correlated with differences in chemical composition. On the other hand, the chemistry of many substances found in muscles is imperfectly known and many constituents have no recognized function.

Muscular tissue of the mammalian skeletal type is composed, as regards main constituents and in round numbers, as follows:

Water	75%
Soluble material (not protein or carbohydrate)	3–5%
Carbohydrate	1%
Fatty acid (minimum)	$\frac{1}{2}$–1%
Protein	18–20%

In many muscles the fat contents will be higher owing to storage fat. *Cholesterol* in small amounts is always present. In visceral vertebrate muscle there is more water and less protein, and as little as 12 per cent. protein in some Invertebrates.

INORGANIC CONSTITUENTS

The inorganic salt content varies in different animals according to the general level of salt content. For instance, the muscles of marine Invertebrates will be isotonic with sea water and have an osmotic pressure about five times that of frog's muscle. As in most tissues, the potassium content of muscle is high and the sodium relatively low. In Table 1 are given some figures for the ash analysis of various vertebrate muscles. They are taken from a review of the chemical composition of muscle by Costantino (1923). In skeletal vertebrate muscles hardly any of the phosphorus is present as inorganic phosphate. Part is ether-soluble, probably *phosphatide* phosphorus; the rest, which is water-soluble, is "organic phosphate", partly phosphate esters. In vertebrate visceral and some invertebrate muscles there is relatively more inorganic phosphate.

Table 1

Inorganic Constituents of Vertebrate Muscle
mg. per 100 gm.

Species	Tissue	K	Na	Ca	Mg	Fe	P	Cl	S	
Ox										
(*Bos taurus*)	Skeletal (Beef steak)	366	65	2	24	25	170	57	187	(1)
	Smooth (*Retractor penis*)	267	109	9	12	2	113	128	217	(2)
Fowl										
(*Gallus indicus*)	Skeletal white	410	89	—	—	—	251	26	—	(3)
	Skeletal red	373	77	—	—	—	241	33	—	(3)
	Stomach	356	72	—	—	—	182	84	—	(3)
Frog										
(*Rana catesbiana*)	Skeletal	350	54	28	30	10	155	66	141	(4)
	Smooth	325	73	4	13	0·7	137	120	161	(4)

References: (1) Katz, 1896; (2) Costantino, 1911; (3) Costantino, 1912; (4) Meigs and Ryan, 1912.

The organic extractives are a miscellaneous group of water-soluble substances mostly of unknown function. Among them *creatine* is conspicuous in vertebrate muscle, which may contain

1 per cent. In the elasmobranch Fishes nearly 2 per cent. of *urea* is present. For the special problems concerned with creatine the reader should consult Hunter's review (1922), and book (1928). It is conceivable that the functions of the living cell call for a certain concentration of diffusible non-electrolytes and that creatine, urea, and the other soluble innocuous organic substances found in muscles fulfil this need. The creatine, however, is partly combined and may have special functions (see p. 11). Among the organic extractives *lactic acid* is sometimes included, but this substance is present during life in minimal quantities, is related to the carbohydrates, and has a special function which will be discussed later.

Inositol is generally present, and may be connected with the carbohydrates (see pp. 50, 51 and also J. Needham, 1926).

CARBOHYDRATES

Of the carbohydrates *glycogen* is the most important and abundant. It has been found in every type of muscle that has been properly investigated. Cases are reported in the literature where glycogen appears to be absent in muscle, but these results are almost certainly due to workers not realizing the rapidity of *post mortem* glycogenolysis, and to the difficulties of analysis where low concentrations are concerned. The amount varies from about 0·1 to 1 per cent. in most muscles. In vertebrate muscle there is a rough correlation between normal glycogen content and degree of muscular activity. In some Invertebrates as in bivalve Molluscs and Nematode Worms considerable quantities (2–3 per cent.) of glycogen are found in muscles and other tissues. The high glycogen content is not connected with excessive muscular activity but is storage material; these animals store glycogen rather than fat and have no specialized tissue or organ for the purpose. Among Crustacea glycogen is stored in the muscles in preparation for the moult and is used up in the process (the muscle tissues as a whole seem to degenerate also). As glycogen is not a well-characterized substance it is difficult to decide whether it is identical throughout the animal kingdom, but Harden and Young (1902) in their very careful work could find no evidence of difference between specimens from yeast, oysters and rabbit's liver.

There is generally found in muscle a small amount, of the order of o·2 per cent., of material which unlike glycogen is soluble in alcohol and gives the reactions of reducing sugars. Some of this is probably free glucose or at any rate is identical with the sugar of the blood. Some of it certainly consists of *hexose esters* of *phosphoric acid*. Embden and Zimmermann (1924) have succeeded in isolating a substance from muscle which appears to be identical with the *hexose diphosphate* obtained from yeast, and also a *monophosphate* (1927). These lower carbohydrates are presumably intermediates in the interconversion of glycogen and lactic acid which occurs during the normal metabolism of the muscle.

A muscle constituent which must be closely concerned with intermediate carbohydrate metabolism is the *phosphagen* of Eggleton and Eggleton (1927). This, as has been mentioned, is a labile compound which breaks down to give creatine and phosphate. It is generally present in the skeletal muscles of Vertebrates, including also *Amphioxus* (Eggleton and Eggleton, 1928). In frog's muscle at rest the creatine is mostly if not entirely combined as phosphagen. In frog's heart there is very little and in stomach muscle none. These facts and others as to its relative amount in different species of animal and types of muscle suggest a connection between speed of contraction and phosphagen content. It has not been found in invertebrate muscle, which also lacks creatine. But Meyerhof (1927) finds a similar substance, probably an *arginine* compound, in crab muscle.

FATS

The fatty materials in muscular tissue present a very difficult analytical problem. There is evidence that fatty material may be stored in certain muscles (Terroine, 1919; Greene, 1926) in the fibres themselves. We must therefore distinguish (*a*) the essential fat of the muscle cell, which is functional and cannot be removed without destroying the cell (*élément constant* of Terroine and his colleagues), (*b*) the storage fat of the cell, which fluctuates according to the food supply of the animal, (*c*) the storage fat of interstitial tissue, which can be distinguished easily histologically but cannot be separated from the muscular tissue for purposes of chemical examination. That there is such a thing as essential fat (*a*) cannot be doubted. In frog's muscles no fatty material can normally be seen microscopically in the fibres or interstitially. But fatty material can always be extracted by fat solvents. More important still it has been found impossible by starvation or other means to reduce the fat content of the muscles of Mammals below a certain minimum (Terroine, 1919, quoted by Leathes and Raper, 1925, p. 96). Normal heart muscle contains only essential fat (a).

It is difficult, if not impossible, to remove the fatty constituents of muscular tissue quantitatively by extraction with fat solvents, so that for quantitative work it is necessary to saponify and estimate the fatty acids obtained. These are highly unsaturated and usually resemble the fatty acids of liver more than those of *adipose* tissue. The saponification method gives no information as to the state of combination of the fatty acids. Direct extraction of frog's muscle, or other tissue in which there is not much storage of fat, with fat solvents yields a material consisting mainly of complex fatty acid esters such as *lecithin* and other *lipins*. There is probably little or no free fatty acid or soap in any tissue. The fatty acid may be assumed to be present in some kind of *ester* form. In mammalian muscle, which stores fat, there is a difference according to the state of nutrition of the animal. In well-fed animals the fatty acid content is high, as is also the ratio *fatty acid/ether-soluble phosphorus*. The value of the ratio is higher than that of lecithin or any known phosphatide, suggesting that the fats may be stored as glycerides. On the other hand, in starvation the total fatty acid sinks, and the ratio *fatty acid/ether-soluble phosphorus* sinks too, to a value lower than any known phosphatide (Mayer and Schaeffer, 1914). The exact significance of this is doubtful because the ether-soluble phosphorus is not necessarily all phosphatide phosphorus.

Cholesterol is always present and is not diminished by starvation (Terroine, 1919, p. 88). Mayer and Schaeffer (1913) find a constant ratio (*lipocytic coefficient*) between the cholesterol and fatty acid content of muscles of any species after starvation and also characteristic ratios for other organs. This ratio *cholesterol/fatty acid* varies with the water content of the tissues and their capacity to imbibe water (Terroine and Weil, 1913). This suggests that cholesterol favours intake of water and fatty acid opposes it.

Clark and Almy (1918) find also in the muscle of fishes which store fat that the water varies inversely as the fatty acid content, while protein remains constant.

For a full discussion of the whole question of the functions of the fatty constituents of organs Leathes and Raper's monograph (1925) should be consulted and also Leathes' Croonian Lecture (1925).

In the normal living muscle fibre of many animals no fat is visible microscopically, so that it must either be emulsified in droplets of sub-microscopic dimensions or (more probably) distributed about the cell substance in films of molecular dimensions. All other ingredients of the cell are water-soluble except for some of the protein and that at least is *hydrophil*, so that it may be assumed on general grounds that one of the functions of the fatty constituents is to confer on the external and internal membranes their peculiar permeabilities and impermeabilities. A mass of protein containing water and salts would be freely permeable to water and all dissolved substances of small molecular weight, with certain limitations in the case of electrolytes imposed by the conditions of the *Donnan Equilibrium*. But a muscle, like other cells, though permeable to water, has specific and variable permeabilities for dissolved substances. It is reasonable to associate these properties with some special variable molecular arrangement of the fatty constituents within the protein framework of the cell.

PROTEINS

This brings us lastly to a consideration of the protein of the muscle, the most abundant and least understood of the solid constituents. The behaviour of the muscle is closely bound up with the nature of its protein, and it is mainly to the proteins that we must look for the specific differences between different animals. Therefore some treatment here of the general character of the proteins is justifiable, though up to the present application of general knowledge to the special problems of muscle physiology cannot be made, except very tentatively. For an excellent monograph on proteins see Lloyd (1926).

The proteins are highly complex substances containing in their molecule the elements C, H, O, N, usually S, very often P, and occasionally other elements. The bulk of the material is built up by the union of some twenty different *amino-acids*, and the behaviour of a protein is essentially that of a complex amino-acid of high molecular weight. Amino-acids are *amphoteric* substances or *ampholytes*, that is they behave as acids in virtue of their *carboxyl* group ($-COOH$), as bases in virtue of their *amino* group ($-NH_2$). In the presence of a strong base they combine as acids to form salts, with a strong acid they combine as bases to form salts. In any solution of an amino-acid there is a region intermediate between alkalinity and acidity when it is not combined with either acid or base (or if combined, not with excess of either). This is the *isoelectric* point or region; so called because on electrolysing the solution at that point the amino-acid does not move to either anode or kathode (or moves equally to both), whereas at a more alkaline reaction it behaves as an acid and moves to the anode, at a more acid reaction it moves to the kathode. In the presence of the acids or bases with which an amino-acid forms soluble salts the isoelectric region is one of minimum solubility and is a critical region for all physical properties. Of the naturally occurring amino-acids some are *monoamino-monocarboxylic* and therefore approximately neutral in behaviour or as strongly acidic as basic. Others are *diamino-monocarboxylic* and predominantly basic; others are *monoamino-dicarboxylic* and predominantly acidic.

Proteins have all the characteristics of the amino-acids by the condensation of which they are formed. It has been shown by Sörensen (1917) and his school, and by many others, that proteins in solution combine with acid and base like simple ampholytes in accordance with the *Law of Mass Action* and the *Ionic Dissociation Theory*. Proteins may contain in the molecule as well as amino-acids, *ammonia* combined as *acid-amide*, carbohydrate groups, inorganic radicles like phosphoric acid, and many other molecular groups.

Among the proteins that have been carefully studied the smallest molecular weight yet found is 34,000, that generally assigned to egg albumin (Sörensen, 1917; Cohn, 1925). Other proteins are known to have molecular weights up to six times as large. Egg albumin is very soluble and forms solutions of low viscosity from which it may be obtained crystalline. Comparing different proteins, increasing molecular weight is accompanied by smaller solubility, greater viscosity of solutions and greater difficulty of crystallization. Many proteins are practically insoluble in the isoelectric region but form soluble salts; some proteins are quite insoluble though they will swell up in water or in aqueous solutions. Solutions of proteins are not as a rule stable but show ageing effects such as gradual change in viscosity. Excess of acid and still more of alkali brings about irreversible changes, probably hydrolytic, slowly at room temperature and very quickly near the boiling point. Many proteins are quickly and irreversibly thrown out of solution by apparently trivial changes in their environment. As the result of heating for a few seconds to 60° C., exposure to sunlight or mere aggregation at a surface, egg albumin is converted into an insoluble form. The change occurs in two stages, the first is an internal change known as *denaturation* for which the presence of water is necessary, the second is a *flocculation* of the denatured material for which the presence of salts is necessary. Denaturation may be defined as an irreversible chemical change whereby a *hydrophil* colloid acquires a character which resembles that of the *hydrophobe* colloids. The change is not necessarily a degradation of the protein molecule (Sörensen, 1924) and does not necessarily involve changes in the acid and base-combining capacities of the protein (Harris, 1923), though such changes may and frequently do accompany denaturation when it is brought about by heat or by acids or alkalis. In common with egg albumin most proteins of the living cell undergo these changes and are grouped together as *heat-coagulable*.

Finally it may be mentioned that many common proteins of animal origin are more acidic than basic and have an isoelectric point in the neighbourhood of pH 5.

Turning now to the muscle proteins themselves. It is possible to squeeze a juice out of muscle tissue which contains some of the protein in solution. By repeated extraction with salt solutions a larger portion can be obtained in solution. These solutions are highly viscous and unstable. Some or all of the protein is readily precipitated and precipitation is usually irreversible. In fact the most striking characteristic of the muscle proteins is the ease and rapidity with which they are denatured and coagulated. Attempts have been made to fractionate the soluble material by fractional heat coagulation and other means. It is difficult to be certain that the various fractions described are anything but stages in the denaturation of the protein or varying mixtures of protein with lipin material.

Generally speaking the most active and powerful muscles, Bird's and Mammal's skeletal muscles, contain the most protein material, and the more sluggish and more feeble mammalian smooth muscles and the muscles of the lower Invertebrates contain the least. This would fit in with the view that one important function of the proteins is to act as buffers for the neutralization of acid. However, Osborne and his fellow-workers (1908, 1909), studying the amino-acid content of the total protein of skeletal muscle, could find little difference between ox, fowl, halibut and the adductor muscles of *Pecten*. But it must be remembered that the method would show only large differences. As between the soluble and insoluble proteins in one kind of muscle no significant difference of amino-acid content has been found, but the proportions are different. In mammalian muscle the skeletal contains proportionally the most soluble proteins, the intestinal muscle the least, while heart muscle is intermediate (Saxl, 1907).

Weber (1925) finds that a soluble protein "myogen" from rabbit's muscle has an isoelectric point at pH 6·3. The insoluble residue has an isoelectric point at pH 5·1–5·2. He shows that this soluble fraction readily forms complexes with salts on the acid side of the isoelectric point, giving an apparent second isoelectric point about pH 5. In spite of this he considers that there is a second soluble protein with an isoelectric point pH 5 which is the "myosin" recognized by von Fürth and other previous workers. In frog's muscle the isoelectric points of soluble (myogen) and insoluble proteins are pH 6·0 and 5·1–5·2 respectively. The difference in isoelectric point of frog's and rabbit's myogen may be of functional significance. Experiments on the region of minimum swelling of frog's muscle tissue point to pH 5·0–6·5 as the isoelectric region (Lloyd, 1917). These results suffice to show that under the conditions in the living muscle the reaction is on the alkaline side of the isoelectric point.

Furusawa and Kerridge (1927, 1, 2), using an electrometric

method developed by Kerridge (1926), have measured the hy-drogen ion concentration of the muscles of a number of animals; skeletal and cardiac muscles of the cat, skeletal muscles of dogfish (*Scyllium canicula*), Crustaceans (*Homarus vulgaris, Eupagurus bernhardus, Maia squinado*), and the adductors of *Pecten oper-cularis*. All these immediately after excision had a pH 7·0–7·1 at room temperature. On standing 20–24 hours for maximum develop-ment of acid the pH was in the neighbourhood of pH 6·3, except for cat's skeletal muscle which was more acid, about pH 6·0. Similar *post mortem* results were obtained with two teleostean fishes (*Conger vulgaris* and *Lophius piscatorius*) and the longi-tudinal body muscles of *Holothuria nigra*. Measurements on frog's skeletal muscle by less reliable methods of Meyerhof and Loh-mann (1926) have given results similar to those on cat's skeletal muscle.

They calculate the true resting value of cat's skeletal and cardiac muscle at body temperature to be pH 7·13 and 7·15 respectively.

The proteins if they are ionized will be anions, that is to say they will behave as acids combined with base. This is an important point, as will be seen later, because recent views on muscular processes turn upon the assumption that the protein material of muscles or some of it behaves as a weak acid combined with a strong base, mainly potassium.

In the main the mechanical properties of the muscles must depend upon the proteins as they are the most abundant solid constituent. The living muscle of a frog is a jelly-like body and is within limits elastic. It is not unlike india-rubber in its tensile properties but less extensible. It does not follow Hooke's law exactly but becomes less extensible the greater the load. It shares with india-rubber the peculiar property of being warmed by stretching and cooled by release to its natural length (Hartree and Hill, 1920). These effects are reversible or at least would be if stretching and releasing were carried out sufficiently slowly. Under ordinary experimental conditions the speed is too great so that the cooling on release is less and the heat on stretching more owing to internal friction in the muscle. The observations are confined to frog's skeletal muscle, but there is no reason to suppose that in this respect other muscles are fundamentally different. We have considered the structure of the muscle as it exists in the living state at rest as nearly as the experimental difficulties enable us to handle the living muscle.

RIGOR MORTIS

The next chapter will deal with the changes that occur in the living muscle on excitation. First it is necessary to consider briefly some changes that occur when a muscle dies. Up to a point the changes are analogous to functional processes of the living tissues, and it is important (but not easy) to distinguish processes which are essentially life processes, from others which are essentially death changes. In this connection it may be well to emphasize that the only convenient criterion of whether a change is a death change or a life change is whether it can be reversed and repeated. Any change brought about under any condition which is completely reversible so that the tissue can be restored to its initial state, and which can be repeatedly brought about in the same muscle will be treated as a *life process* or as *normal* or *functional*. Any change which is not reversible or cannot be repeated must be open to suspicion as a *death change* or as *abnormal*. These words are not used in any mystical sense but purely empirically. Difficulties in interpretation arise in practice on account of the difficulty of keeping a tissue in a steady state under conditions of experiment. It is liable all the time to be *in articulo mortis*.

Vertebrate skeletal muscle, and to a less extent vertebrate heart muscle, when it dies undergoes a process known as *rigor mortis* which terminates in shortening of the muscle, and development of rigidity in the animal. The most marked chemical process accompanying it is the breakdown of *glycogen* to *lactic acid*. This is not in itself a death change, but in death much more lactic acid is produced than ever appears during life. This glycogenolysis is controlled by (1) the general speed of chemical change in the muscle—its time scale, (2) the amount of glycogen or other carbohydrate present which can give rise to lactic acid, (3) the buffering capacity of the tissues, since glycogenolysis is a self-inhibiting reaction in the sense that increasing acidity slows it down (see Meyerhof, 1921; Hines, Katz and Long, 1925; Ritchie, 1927). The production of acid causes a measurable increase in the acidity of the whole tissue. The shortening and hardening of the tissue appears to be in part the result of this and to be analogous to the precipitation of protein, or in other words the change in reaction of the tissue has brought the proteins nearer their isoelectric point and diminished the stability of their solution. Though a muscle shortens in rigor and on stimulation during life it does not follow that the two processes are similar, as is sometimes assumed, particularly when we consider that some muscles do not go into rigor at all. A muscle under no load shortens more in rigor than under maximum stimulation in life but produces less tension. The mechanical properties of the tissue in rigor are different, too; it is less elastic and more easily passively deformed.

Certain reagents, such as chloroform, ether, and also heat produce a breakdown of structure, a denaturation of muscle proteins and a consequent glycogenolysis which also results in shortening. These types of change, *chloroform* or *heat rigor*, ought to be distinguished from natural rigor which involves less drastic changes and no obvious denaturing of proteins, i.e. some remain soluble in salt solutions. The development of *rigor mortis* is characteristic of vertebrate striped muscle (skeletal or heart). It is absent in the smooth muscle of vertebrate viscera (Evans, 1926, 2). Among Invertebrates it is seldom conspicuous even in skeletal muscle, and may be absent altogether. The prolonged contraction observed by Mangold (1926) in isolated strips of rat's stomach can hardly be considered to be rigor, because it set in and passed off quicker than rigor of skeletal muscle, and in some cases the muscle was still contracting rhythmically at the height of the curve of contraction.

In crustacean muscle according to Hoet and Kerridge (1926), confirming Claude Bernard's observations, *rigor mortis* is most marked just after the moult, when the muscle contains little glycogen, and occurs without the development of acidity. Hoet and Marks (1926) had shown already that the same thing is true of mammalian muscle in animals deprived of muscle glycogen by *insulin* or other means. Therefore there is a type of *rigor mortis* which is due simply to lack of glycogen or to some related change and not to accumulation of lactic acid. Evans (1926, 1) has found that *caffeine*, which causes rapid glycogenolysis and rigor in vertebrate skeletal muscle, actually inhibits glycogenolysis in smooth muscle. He correlates this with Secher's (1914) histological observations that caffeine and other *purin bases* produce a break-down of structure of vertebrate skeletal but not of invertebrate muscle. In frog's skeletal muscle chloroform brings about a breakdown of structure causing a shortening which precedes the production of lactic acid, though the acid in its turn when it appears causes further shortening (Ritchie, 1925). The effect of chloroform is more violent than that of caffeine, as if concentrated enough it denatures the proteins of any tissue.

Another explanation of the absence of rigor in mammalian smooth muscle may be based on Furusawa and Kerridge's (1927, 2) *buffer curves* for cat's uterus muscle. It seems likely that the maximum lactic acid produced by the tissue will not bring its reaction to the isoelectric region of the proteins, as happens in skeletal muscle. Before this explanation can be applied generally more information on the isoelectric points of muscle proteins is needed.

In the ordinary form of *rigor mortis* Meyerhof and Himwich (1924) found a close relationship between the amount of shortening and the amount of lactic acid formed. We may conclude therefore that there are three factors concerned: (1) a certain instability of cell structure, (2) increase of that instability by removal of glycogen or possibly concomitant changes in other carbohydrates, (3) the production of lactic acid, which by bringing the proteins nearer their isoelectric point causes diminished stability in another way. If the breakdown of structure is brought about rapidly by chloro-form or caffeine it is visible microscopically, otherwise there may be no visible change except shortening. Bringing proteins nearer their isoelectric point generally removes water from them. The removal of water from one set of cell constituents may contribute to the process. It is possible that glycogen in the muscle is asso-ciated with the protein structure concerned, and that the effect of diminishing the amount of this hydrophil colloidal substance may be simply the removal of water from that part of the structure.

A more extensive study of the death changes in various types of muscle might reveal the nature of the instability that appears as rigor and show its functional significance. As far as the evidence goes at present it looks as though the muscles that performed the heaviest duties were the least stable in this sense. It is noteworthy that they are also the muscles that contain the largest amount of soluble protein.

In addition to the changes described there is a small *post mortem* production of ammonia in muscle or at any rate of a volatile base which gives the Nessler colour reaction. Gad-Andersen (1919) attributed it to a tissue *urease* and certainly the urea content of the tissues is sufficient to supply the ammonia liberated. Parnas and Mozolowski (1927) however find practically no breakdown of urea. They have studied the ammonia production extensively and find it is produced by injury, by all the forms of "rigor" and also in smaller quantity by stimulation. The injury production is to some extent inhibited by treatment with alkaline borate solution. Ammonia is produced in skeletal muscles of Fish, Amphibians, Reptiles, Birds, and Mammals. The largest amount, about 8 mg. per cent. in Fish, smaller amounts, 3–5 mg. per cent. in higher animals. Smaller quantities still are formed in heart and stomach muscle. The origin and significance of this ammonia production must be considered obscure. The quantities formed are small relatively to the quantities of lactic acid formed and the change of hydrogen ion concentration produced by the liberation of ammonia, even assuming its precursor is unionized, is probably negligible in comparison.

REFERENCES

CLARK and ALMY (1918). *J. Biol. Chem.* **33**, 483.

COBB, S. (1925). *Physiol. Rev.* **5**, 518.

COHN, E. J. (1925). *Ibid.* **5**, 349.

COSTANTINO (1911). *Biochem. Zeit.* **37**, 52.

— (1912). *Ibid.* **43**, 165.

— (1923). *Arch. di Sci. Biol.* **4**, 448.

EGGLETON and EGGLETON (1927). *Biochem. J.* **21**, 190; *J. Chem. and Ind.* **46**, 485.

— — (1928). *J. Physiol.* **65**, 15.

EMBDEN and ZIMMERMANN (1924). *Zeit. Physiol. Chem.* **141**, 225.

— — (1927). *Ibid.* **167**, 114, 137.

EVANS, C. L. (1926, 1). *Physiol. Rev.* **6**, 358.

— (1926, 2). *Biochem. J.* **20**, 893.

FULTON, J. F. (1926). *Muscular Contraction.* London and Baltimore.

FURUSAWA and KERRIDGE (1927, 1) *J. Marine Biol. Ass.* **14**, 657.

FURUSAWA and KERRIDGE (1927, 2). *J. Physiol.* 63, 33.
GAD-ANDERSEN (1919). *J. Biol. Chem.* 39, 267.
GREENE (1926). *Physiol. Rev.* 6, 229.
HARDEN and YOUNG (1902). *J. Chem. Soc.* 81, 1224.
HARRIS (1923). *Proc. Roy. Soc.* B, 94, 426.
HARTREE and HILL (1920). *Phil. Trans.* B, 210, 153.
HAYCRAFT (1891). *Proc. Roy. Soc.* 49, 287.
HILL, A. V. (1926). *Muscular Activity.* Baltimore.
HINES (1927). *Quart. Rev. Biol.* 2, 149.
HINES, KATZ and LONG (1925). *Proc. Roy. Soc.* B, 99, 8, 20.
HOET and KERRIDGE (1926). *Proc. Roy. Soc.* B, 100, 116.
HOET and MARKS (1926). *Ibid.* 100, 72.
HUNTER (1922). *Physiol. Rev.* 2, 586.
— (1928). *Creatine and Creatinine.* London
KATZ (1896). *Pflüg. Arch.* 63, 1.
KERRIDGE (1926). *J. Sci. Inst.* 3, 404.
KNOLL (1891). *Denkschr. Akad. Wien*, 58, 633.
LEATHES (1925). *Lancet*, pp. 803, 853, 957, 1019.
LEATHES and RAPER (1925). *The Fats.* London.
LLOYD, D. J. (1917). *Proc. Roy. Soc.* B, 89, 277.
— (1926). *Chemistry of the Proteins.* London.
MANGOLD (1926). *Ergeb. d. Physiol.* 25, 46.
MAYER, A. and SCHAEFFER, G. (1913). *J. Phys. et Path. gén.* 15, 510, 535,
    773, 984; 16, 1, 23.
— — (1914). *J. Phys. et Path. gén.* 16, 204; *Comptes Rendus Acad. Sci.*
    159, 102.
MEIGS and RYAN (1912). *J. Biol. Chem.* 11, 401.
MEYERHOF (1921). *Pflüg. Arch.* 188, 114.
— (1927). *Naturwiss.* 15, part 32.
MEYERHOF and HIMWICH (1924). *Pflüg. Arch.* 205, 415.
MEYERHOF and LOHMANN (1926). *Biochem. Zeit.* 168, 128.
NEEDHAM, D. M. (1926). *Physiol. Rev.* 6, 1.
NEEDHAM, J. (1926). *Ergeb. Physiol.* 25, 1.
OSBORNE *et al.* (1908–1909). *Am. J. Physiol.* 22, 433; 23, 81; 24, 161, 437.
PARNAS and MOZOLOWSKI (1927). *Biochem. Zeit.* 184, 399.
RITCHIE (1925). *J. Physiol.* 59, lxxxii.
— (1927). *Brit. J. Exp. Biol.* 4, 327.
SAXL (1907). *Hofmeisters Beitr.* 9, 1.
SCHAFER, Sir E. S. (1912). *Textbook of Microscopic Anatomy.* London.
SECHER (1914). *Arch. exp. Path. Pharm.* 77, 83.
SÖRENSEN (1917). *Comptes Rendus Lab. Carl.* 12.
— (1924). *Ibid.* 15, No. 9.
TERROINE (1919). *Physiologie des Substances grasses.* Paris.
TERROINE and WEIL (1913). *J. Phys. et Path. gén.* 15, 549.
WEBER (1925). *Biochem. Zeit.* 158, 443, 473.
WINTON (1927). *J. Physiol.* 61, 368.

Chapter II

CHEMICAL CHANGES IN ACTIVE MUSCLE

§ I. *FROG'S MUSCLE*

This chapter will be devoted to a treatment of the chemical events that occur in a muscle when it is stimulated. The skeletal muscle of the frog has been so much more thoroughly studied than any other that we must begin with it and afterwards compare or contrast other muscles wherever anything happens to be known about them. As there are several readily accessible monographs on the processes in frog's muscle to which the reader may be referred for details, the matter will be dealt with briefly (Evans, 1926; Hill, 1916, 1922, 1925, 1926; Meyerhof, 1924; Hill and Meyerhof, 1923). For particulars as to ordinary experimental treatment of isolated tissues, methods of stimulating and recording the response of muscles, reference should be made to the standard textbooks of Physiology. It is perhaps necessary to warn the reader that many of the published records of the response of muscles to stimulation depend much more upon the mechanical properties of the levers to which the muscle is attached than upon any inherent property of the muscle itself. In fact the apparently simple task of making an accurate record of the mechanical changes in a stimulated muscle is, of its kind, one of the most baffling the physiologist is faced with. For an account of recent technical improvements in mechanical recording see Fulton (1926, pp. 85–176). The skeletal muscle of a Vertebrate isolated from the body of the animal continues to respire as long as it remains alive, but under normal conditions has no other sensible activity when left to itself. When it is stimulated electrically or otherwise it gives a response which under suitable conditions can be made substantially identical with the natural response during life. Accompanying the response are a number of changes, mechanical, thermal, chemical and electrical. Some of these are directly concerned with the change that is specific for muscular tissue, namely the mechanical; others are changes common to all excitable cells.

It is important to distinguish as far as possible between the two kinds of process. We may assume with a fair degree of confidence that the *electrical change*, which accompanies muscular activity, is not merely part of the system of processes to which the *mechanical change* belongs, since all types of excitable tissue give an electrical change. So that an *excitatory process* which muscle shares with nerve or gland cells is to be distinguished from the *contractile process* proper.

EXCITATION AND CONTRACTION

Earlier work of Mines (1912) and others appeared to show that in the frog's heart perfused with calcium-free solutions the mechanical response is more reduced than the electrical. If this were right it would show clearly that the two processes were distinct. However, the experiments purporting to show this have been criticized on the grounds that the mechanical recording was not sensitive enough. Einthoven and Hugenholtz (1921) and Arbeiter (1921), using an improved technique, found that when the heart was perfused with excess of potassium or deficiency of calcium the mechanical and electrical responses diminished and disappeared together. Similarly they reappeared and increased together on perfusing with normal fluid. These experiments show that the two phenomena are not separable by this means and hence are intimately connected, not necessarily that they are manifestations of the same process. MacSwiney and Mucklow (1922), using frog's skeletal muscle, measured the total amount of electricity produced in a short tetanus by charging up a condenser from the muscle and measuring the discharge afterwards. The electrical charge was found to increase in proportion to the duration of stimulation. With stimuli of frequency up to that required to produce complete fusion of contractions, the electrical charge was proportional to frequency. As the heat production, which is a measure of the total mechanical change, does not show this simple linear relationship to duration (Hartree and Hill, 1921, 1) or frequency of stimulation (Hill, 1914), the electrical and mechanical changes are not simply different manifestations of the same process. Bishop and Gilson (1927), as the result of very careful work on frog's skeletal muscle, distinguish two parts of the electrical response, the first a very rapid and large effect, and the second slower and less conspicuous. By treating muscles with water, which destroys contractility without entirely destroying the capacity of the tissue to conduct excitation, the second part of the change is abolished but the first only slightly diminished. Therefore it would appear that the first electrical change is at least partially separable, the second inseparable from the contractile process. If the electrical change, at any rate the first part of it, is a measure of the excitatory process E it will be related to the mechanical process M as cause to effect. Consequently any change in E, such as we may suppose produced by excess of potassium or deficiency of calcium in frog's heart, would affect M concomitantly. But clearly it may be possible to alter M without producing an equal change in E, as in Bishop and Gilson's experiment.

Lastly, the chemical processes accompanying the passage of an impulse in a nerve fibre, which we may regard as a purely excitatory process, are found to be noticeably different from the chemical processes underlying muscular contraction which are about to be described (Gerard, 1927). In what follows attention will be directed to these only. The mechanical, chemical and thermal changes will be treated as manifestations of one fundamental process or system of processes, the contractile process. The electrical changes will not be further considered except incidentally.

RESTING METABOLISM

Resting muscle has a definite *oxygen intake* and *carbon dioxide output* which is constant at any given temperature, as long as the muscle is alive and uninjured, and which increases with rise of temperature. In isolated tissues the metabolic processes at rest differ from the active processes in quantity rather than in quality; that is to say, the final result of stimulation is to accelerate chemical changes already taking place at a slow constant rate. Burn and Dale's (1924) experiments on the respiration of resting mammalian muscle in animals which had been eviscerated but had their blood supply intact make it probable that the processes in the animal are similar to those when the tissue is isolated. The qualitative similarity between resting and active metabolism probably extends also to processes which precede intake of oxygen and output of CO_2, such as the breakdown of glycogen to lactic acid. What may be the function of the resting metabolism is hard to say except that the living state is one of instability and that work must be done by the cell on its own substance to maintain it. It is always assumed in studies on muscular activity that the metabolism of the active muscle is superimposed on a constant resting metabolism which can be measured before and after. To arrive at the gaseous exchange of activity the resting gas exchange is subtracted from the gross exchange observed during activity. The validity of the assumption that there is a constant *resting* process underlying the *active* processes cannot be tested directly, but the results obtained on this assumption are more reasonable and regular than they would be if it was assumed that the resting metabolism ceased during activity. The same procedure is adopted in experiments on isolated tissues and on the whole animal. In the latter case the "resting metabolism" includes the metabolism of all organs while the subject is motionless and using his muscles as little as possible. It includes of course metabolism that is not muscular, and is to some extent qualitatively different. The matter is mentioned here so that in what follows it may be understood that in speaking of the gaseous exchange of muscular activity it is always the nett exchange after deducting for resting metabolism that is referred to.

CO_2 OUTPUT AND LACTIC ACID PRODUCTION

It has been known for a long time that the rate of respiration of active muscle was greater than that of resting, and that a muscle could function for a time without oxygen, although it became fatigued quickly, and could recover from fatigue only with a supply of oxygen. During the latter part of the nineteenth century the significance of these and other relevant facts was obscured and the progress of knowledge hindered by the general acceptance of the *Biogen Theory*. According to this theory the fundamental change on stimulation of a muscle was the breakdown of a hypothetical unstable giant molecule which constituted the living substance. Naturally such a substance is not amenable to ordinary experimental investigations. Recent advances in knowledge have been the result of substituting for the *ignis fatuus* of biogen the investigation of familiar and relatively simple chemical reactions. The first important step was made by Fletcher (1898) who showed with isolated frog's muscles (1) that resting muscles in oxygen have a steady output of CO_2, (2) that moderate stimulation in an atmosphere of nitrogen produces no increase in CO_2 output, (3) that in nitrogen prolonged stimulation to fatigue, i.e. slow or incomplete relaxation and diminished excitability, produced a small increase in CO_2 output. This third effect was attributed by him to the displacement of CO_2 from bicarbonate by increasing acidity of the tissue, an assumption confirmed by all later workers. The increased acidity in fatigue had been observed many years before by Claude Bernard. Fletcher showed also (4) that a muscle stimulated in oxygen gave an increased CO_2 output and did not become fatigued, and (5) that a muscle fatigued in nitrogen recovered in oxygen, giving out excess CO_2 in the process. These observations sufficed to explode the Biogen Theory, and supplied a basis of correct experimental data on which subsequent work has been built. They showed that oxygen intake and CO_2 output were concerned in the same process, which was essentially a *recovery process* and no part of the *contractile process* proper. Biogen had been assumed to contain a supply of combined oxygen, like guncotton or T.N.T., and to break down, giving rise among other things to CO_2, so that on this theory intake of oxygen and output of CO_2 belonged to different parts of the cycle of changes. There would be no need at the present time to mention a theory which is (or ought to be) extinct, were it not an excellent example of the obstructive type of hypothesis which is liable to hinder investigation. The next important step was made by Fletcher and Hopkins (1907, 1917) in their classical work on lactic acid production in frog's muscle. Previous attempts to correlate muscular processes with the lactic acid content had failed, because, as Fletcher and Hopkins showed, the handling of the muscle in the first stages of chemical analysis is liable in itself to lead to lactic acid production. They found that by rapid chopping and grinding under alcohol cooled in freezing mixture they could kill and extract the tissue with a minimum production of lactic acid. Their work demonstrated (1) that resting muscle contains very little lactic acid, about as much as blood does, (2) that stimulation in nitrogen causes an increase up to a maximum when the muscle becomes fatigued and inexcitable. In oxygen they found (3) that lactic acid does not accumulate with stimulation and that the acid present in fatigued muscle disappears. (4) Muscle repeatedly fatigued and allowed to recover can still develop as much lactic acid as at first. (5) Injury to the tissue or death changes result in lactic acid formation. In *rigor mortis* the muscle contains about twice as much as can be produced by stimulation.

These results confirmed Fletcher's conclusion that the contractile process was essentially *anaerobic*, and in addition showed that the lactic acid was part of the contractile mechanism proper, and is removed in the oxidative recovery process. As to whether the lactic acid is oxidized or whether it is reconverted to the precursor from which it originated by means of the oxidation of something else could not be determined. However, result (4) favoured the suggestion that it was reconverted.

LACTIC ACID AND GLYCOGEN

The part played by lactic acid has been further elucidated by the admirable work of Meyerhof (1919–). He has shown (1) that when lactic acid is formed anaerobically in muscle an equivalent quantity of *glycogen* disappears. Taking glycogen as a polymer of *glucose*, 1 gm. of $(C_6H_{12}O_6)_n$ or 0·9 gm. of the anhydride $(C_6H_{10}O_5)_n$ yields 1 gm. of $(C_3H_6O_3)$ lactic acid. The reaction is essentially a molecular rearrangement and involves neither oxygen nor CO_2. No other significant changes in the carbohydrates of frog's or rat's skeletal muscle have been observed. That this reaction is the essential reaction of the contractile process is shown by (2) the fact that the total tension which the muscle develops is proportional to the amount of lactic acid formed and the ratio is not altered by change of temperature. A similar proportionality has been found in heart muscle (Redfield and Medearis, 1926). In the presence of oxygen he found (3) that as the lactic acid disappears glycogen reappears, not in equivalent amount but to the extent of about three-quarters of the lactic acid. (4) The respiratory exchange during the process accounts for the remaining quarter of the lactic acid. That is to say, oxygen intake and CO_2

output are what would be calculated for the combustion of this amount of lactic acid or its equivalent of carbohydrate.

The data enumerated above have all been got on frog's muscle, but several of the chief experiments have been repeated with mammalian muscle and show no noticeable difference beyond that due to differences in body temperature and the greater difficulty of handling the less robust mammalian tissue.

HEAT MEASUREMENTS

Ordinary chemical methods by themselves can probably not take us much farther; they are somewhat clumsy and inaccurate; considerable quantities of tissue are needed and must be destroyed for each observation. A more refined and subtle method of investigation is provided by the measurement of heat production. By the use of a small thermopile with a large number of metallic junctions made to fit over the muscle, and the use of delicate electrical recording instruments, it is possible theoretically to attain any desired sensitivity. In practice, to measure changes of temperature of the order of $0.001°$ C. requires great care and skill. It is notorious that magnifying the sensitivity of an instrument increases the errors and the difficulty of manipulation. Electrical instruments particularly may prove a trap to the unwary. The successful application of calorimetric methods to the study of muscle physiology has been the work of A.V. Hill (1912), and recent improvements are due to him and his associates (Hartree and Hill, 1921; Fenn, 1924; for full references consult the reviews cited at the end of the chapter).

The great advantage of calorimetric over chemical methods is not so much their greater delicacy, though that is important, as that a series of measurements of heat production of an individual muscle can be made without interfering with or injuring it in any way. On the other hand, without a knowledge of the chemical mechanism at work the heat production records would be written in a cipher without a key.

Most heat measurements have been made on muscles under *isometric* conditions. That is, the muscle has been held at a constant length. Under these conditions no external work is performed because the point of application of the force is not moved and all the energy of the contractile process appears as heat.

Moreover, the internal mechanical changes are reduced to a minimum and are presumably constant for any given tension developed. The mechanical effect of the contractile process is measured by the tension developed.

The main results of this work of Hill and his collaborators are: *in the absence of oxygen* (1) heat is produced concurrently with the contraction and relaxation of a muscle in a single twitch (*initial heat*); (2) the amount and the time course of production of the initial heat are unaffected by the presence or absence of oxygen, taking the strictest possible precautions to eliminate oxygen; (3) the ratio of the initial heat to the tension developed is a constant independent of temperature; (4) there may be a small delayed heat production (*delayed anaerobic heat*) which is not due to oxidation but to some process not well understood (Furusawa and Hartree, 1926). *In the presence of oxygen* there is (5) a second output of heat lasting a considerable time and one and a half times as large as the initial heat (*delayed oxidative heat*) which is clearly produced in the oxidative recovery process.

BALANCE SHEET OF ENERGY EXCHANGE

Meyerhof had found the ratio lactic acid/tension to be constant and Hill the ratio tension/heat. It was to be expected therefore that the ratio heat/lactic acid would be constant. This has been found to be the case first by Hill and Peters, later by Meyerhof. This constant is an important one because it enables the results of calorimetric and chemical observations to be correlated, and the complete balance sheet of energy exchange to be made out. The most recent figure given by Meyerhof (1924) is 390 calories per gram of lactic acid liberated. The exact determination of the ratio is a matter of considerable difficulty and in different series of experiments somewhat different results have been obtained. The determination has been done with frog's muscle only. If we know the heats of combustion of glycogen and lactic acid we can calculate the heat of conversion of the one to the other. In determining the heat of combustion of glycogen the chief difficulty is to obtain a substance free of impurities and of definite composition, in particular of a definite degree of hydration. A small correction has to be made too for the heat of wetting and solution, as the combustion is done on the solid, whereas in the muscle the glycogen is dissolved. Using glycogen from *Mytilus edulis*, a convenient source, Slater has successfully carried out this determination. Meyerhof has obtained rather different results, but as his glycogen contained a considerable amount of ash Slater's figures are probably better. Meyerhof also found a slight difference between glycogens from different animals but the differences are hardly outside the experimental error, taking into account the difficulties of purification. The heat of combustion of lactic acid must be determined indirectly through its salts. This has been done by Meyerhof, who has also found the necessary corrections for solution and dilution. The results are:

Heat of combustion of dissolved glycogen per gram $C_6H_{12}O_6$ (Slater, 1924)	3836 cal.
Heat of combustion of dilute lactic acid per gram $C_3H_6O_3$ (Meyerhof, 1922)	3601 cal.
Difference	235 cal.

This heat is considerably less than the heat developed in muscle, so that another reaction must occur. Muscle tissue is normally neutral or faintly alkaline and is not appreciably changed by the lactic acid liberated in moderate activity, so that the lactic acid cannot remain free but must be neutralized. Neutralization by free alkali would liberate a large quantity of heat, but there is no free alkali present in muscle, i.e. the hydroxyl ion concentration is small even compared to the lactic acid concentration due to a single twitch. Meyerhof found that if lactic acid is neutralized by "buffer salts", such as phosphates or carbonates, only about 19 cal. of heat were liberated per gram of lactic acid. Neutralization by alkali salts of proteins produced very much more. If small quantities of lactic acid are added to muscle proteins when present as alkali salts at a reaction about pH 8 there is little change in reaction but heat production equal to 140 cal. per gram of acid. In other words, the proteins have a large negative heat of ionization. If we assume that this occurs in muscle we then have the relation:

Heat of conversion of glycogen to lactic acid	235 cal. per gram.	
Heat of neutralization by protein	140 cal	,,
Total calculated heat	375 cal.	,,
Total observed heat	390 cal.	,,

The difference is well within the experimental error which must be of the order of 20 cal. per gram in observed and calculated heats. The calculation shows that it is unlikely that any other reaction of importance from the point of view of energy exchange is going on. The breakdown of glycogen to lactic acid may involve a number of side reactions but they are probably nearly isothermic. Strictly intermediate reactions will not result in any nett loss or gain of heat. Furusawa and Kerridge (1927, 2) have calculated the increase in lactic acid in cat's heart and skeletal muscle in fatigue and in rigor from the titration curves of the tissues. The results agree well with the directly determined lactic acid. This provides independent evidence that it is the chief factor in altering the acid-base equilibrium in the tissue.

Meyerhof (1927) gives somewhat different figures for the energy balance sheet of muscle. He takes a lower value for the heat of conversion of glycogen to lactic acid (see p. 31) and considers that part only of the acid is neutralized by protein. His estimate leaves the calculated heat 120 cal. lower than the observed heat. He proposes to make good the deficiency by means of the heat of hydrolysis of phosphagen (see p. 11) which appears to be of the right order. It must be mentioned also that the figure for the observed heat may be too high, as it will include in addition to the *initial heat* the *delayed anaerobic heat* (see p. 30) which may not be part of the contractile process proper. The balance sheet has evidently not yet reached its final form.

INTERMEDIATE METABOLISM

Hartree and Hill (1921, 1) have found, with repeated stimulation of frog's muscle, that heat is first of all liberated at a greater rate than subsequently, when the rate becomes constant to maintain a steady tension. They infer that there is a small "ready store" of material producing lactic acid and that the subsequent rate of heat production is governed by a relatively slow reaction converting glycogen to the intermediate form, the "ready store". The "ready store" may be a labile form of sugar, possibly a phosphoric ester, but there is as yet no reliable evidence on the point; indeed, it is doubtful if present day chemical methods are refined enough to identify it.

There is a considerable amount of organic phosphate in the skeletal muscle which readily breaks down on stimulation to fatigue and under other conditions, but this appears to be a creatine compound and its part in the contractile mechanism is not yet clear (Eggleton and Eggleton, 1927; 1928, 1; and Fiske and Subbarrow, 1927). Equally obscure is the part played by the hexose phosphates isolated by Embden and Zimmermann (1924 and 1927). If the analogy of yeast fermentation can be safely followed, we should expect the hexose diphosphate to be formed in a coupled reaction concurrently with the formation of lactic acid, and from a common precursor probably the monophosphate, and we should not expect it to be itself the direct antecedent (Meyerhof, 1924, Review, p. 47).

As the work of Embden and his school on the chemical changes in muscle and the part played by "lactacidogen" has received considerable attention, something must be said about it; but it is difficult to treat seriously views which appear to be based upon a misconception of the nature of the problem and to be supported for the most part by errors in technique. The Hill-Meyerhof theory expounded in these pages rests upon the discovery that anaerobically heat, tension, and lactic acid production all vary together. An alternative theory or destructive criticism of this theory is worth nothing unless accompanied by evidence that these experimental findings are incorrect, or else that there is some fourth factor which has been neglected and which is a function of the tension developed by a muscle. As the Embden school have not made tension or heat measurements they have obtained no fresh evidence as to the mechanism of muscular contraction. Their experimental findings can be expected to throw light only upon the intermediate carbohydrate metabolism of muscle or upon changes accompanying activity which result in no great liberation of energy and are therefore of minor importance. However, a large part of the chemical evidence that has been put forward rests upon determinations of changes in the inorganic phosphate content of muscle, supposed to reflect changes in a hexose phosphate ester. The discovery by Eggleton and Eggleton (1927) that there is in muscle a very labile organic phosphate, which is not a hexose compound, and the presence of which was not suspected, invalidates much of this evidence. On the other hand, it must be admitted that the work has focussed attention on the phosphate compounds as intermediates in muscle processes, even though the part they play has not yet been elucidated. For a recent account of his views see Embden (1927).

It will be seen from the Tables on p. 49 that there is some tendency for reducing substances soluble in aqueous alcohol to accumulate when glycogen breaks down. Macleod and Simpson (1927) find that glycogen may disappear faster than lactic acid appears, and Olmstead and Harvey (1927) that frog's skeletal muscle which is practically free of glycogen after *insulin* convulsions can still contract normally. All these results suggest that there are intermediates between glycogen and lactic acid which need investigation.

The work of Eggleton and Eggleton (1928, 1) on the *creatine-phosphoric* acid compound of vertebrate skeletal muscle (phosphagen) is of interest in this connection. They find that the compound breaks down on stimulation or in resting anaerobic muscle. In oxygen it is resynthesized. This breakdown and resynthesis are both more rapid than lactic formation and removal. In resting anaerobic muscle the inorganic phosphate liberated is equivalent to the phosphagen broken down, in stimulated muscle it is less, due probably to esterification of the phosphate. If, as seems likely, phosphate esters are concerned in intermediate carbohydrate metabolism, the phosphagen probably supplies the phosphate in a form available for esterification.

The metabolism of carbohydrates in the muscle cannot be entirely divorced from the general carbohydrate metabolism of the body. The muscle gets its supply of carbohydrate from the sugar of the blood which it synthesizes to glycogen. In mammals diminution of the blood sugar does not in itself cause breakdown of the muscle glycogen to replenish the general supply (Soskin, 1927). It would appear that the muscle glycogen is part of the muscle machinery and once formed goes to lactic acid only. This irreversibility of the muscle processes is in strong contrast with the reversible change Blood sugar $\rightleftharpoons$ Glycogen in the liver. The carbohydrate metabolism of invertebrates probably follows a different course.

It is now well established that insulin, the hormone of the *islet* tissue of the pancreas, promotes the formation of muscle glycogen from blood sugar (Best, Hoet and Marks, 1926). Whether it has any other effect on muscle is not clear. Certainly there is no indication of any abnormality in the working of the muscle of depancreatized animals as long as the muscles contain glycogen. For a general discussion of carbohydrate metabolism the reader is referred to Macleod (1925).

RELAXATION PROCESS

Meyerhof concludes that on stimulation lactic acid is produced in the neighbourhood of some specific structure (*Verkürzungsort*) which is altered so as to produce the muscular contraction. After a short interval of time the free acid is neutralized (at the *Ermüdungsort*) by the "buffers" of the tissue and the muscle relaxes. Hartree and Hill, from their heat measurements, find that heat is evolved during relaxation as well as during contraction, and that the relaxation process has a temperature coefficient of a size characteristic of chemical reactions and differing somewhat from the temperature coefficient of the contraction process.

When a muscle has been stimulated anaerobically and the contractile process is finished, the chemical change consists in the disappearance of a little glycogen, the appearance of a corresponding amount of potassium lactate in solution and the de-ionization of some protein. In a single twitch of frog's muscle the amount of glycogen disappearing can be calculated to be of the order of 0·001 per cent. The process is brought to a stop on repeated stimulation, when 200–400 times as much acid has appeared, by the increasing acidity of the tissue as the buffer substances get used up.

OXIDATIVE RECOVERY

In the presence of oxygen combustion takes place, a portion of the lactic acid present or an equivalent amount of glycogen is oxidized and part of the energy liberated goes to reverse the exothermic reaction of the contractile process. Glycogen is reformed and protein re-ionized. So that finally at the end of the oxidative recovery process there is no change but the disappearance of some glycogen. The total heat obtained in the whole cycle of operations represents therefore the heat of combustion of the carbohydrate that has gone, and the "efficiency" of the oxidative recovery process is given by the ratio of this heat (870 cal. per gram of lactic acid formed) to the heat of combustion of that amount of glycogen which gave rise to lactic acid (3836 cal. per gram); that is 1/4·4. To calculate the heat production in terms of lactic acid from the heat measurements of Hill and his associates it is necessary to apply the heat/lactic acid ratio obtained by Meyerhof.

The latest results of heat measurements are summarized by Furusawa and Hartree (1926) as follows:

Process	Relative heat	Absolute cal./gm. lactic acid
Total anaerobic	1·12	390
Initial anaerobic	1·00	348
Delayed anaerobic	0·12	42
Delayed oxidative	1·50	522
Total oxidative	2·50	870

The figures in this "balance sheet" are to be considered as tentative only. The value 390 cal. per gram of lactic acid, from which the other absolute values are calculated, may be in error by 10 per cent. or so and the amount and significance of the delayed anaerobic heat is still not settled definitely. (See Appendix.)

The "efficiency" of the oxidative process has been redetermined recently by Meyerhof and Schulz (1927) by the direct chemical method. They measured (a) the ratio of tension developed to lactic acid produced anaerobically, (b) the ratio of tension developed to oxygen consumed aerobically. From these two values the ratio lactic acid oxidized/lactic acid developed is calculated. The mean is 1/4·7 with considerable variation in different experiments.

CONCLUSIONS

We may conclude then that in frog's skeletal muscle the essential chemical changes during activity are:

(1) Conversion of *glycogen* to *lactic acid*, which occurs on *contraction*.

(2) Neutralization of *lactic acid* by protein buffers, which occurs on *relaxation*.

(3) Partial resynthesis of *glycogen* from *lactic acid* at the expense of the combustion of the residue or an equivalent amount of carbohydrate, which occurs in *oxidative recovery*.

That other changes occur is practically certain but they are likely to be slow, on a small scale, or else intermediate. If they are intermediate they will not affect the nett energy exchange. There can be no doubt that phosphates and carbohydrates simpler than glycogen play an important rôle in the intermediate stages, but the evidence on these points is still confused and incomplete. It is better therefore to leave it out of account at the present.

We have now two problems to consider. (1) Can the scheme which holds good for frog's skeletal muscle be applied generally to all types of muscle? (2) Assuming the answer to (1) to be affirmative, as regards the usually occurring process, is it possible for any muscle under any conditions to make use of a different mechanism; more especially, can the energy be supplied by the combustion of substances other than carbohydrates, say fats? Taking these questions separately in order.

The first question resolves itself into four separate questions: (a) Does lactic acid appear in all types of muscle under conditions which make it appear in frog's muscle? (b) What are the quantitative differences between different muscles? (c) Is the lactic acid formed from glycogen? (d) Is it resynthesized to glycogen in an oxidative recovery process?

§ II. *OTHER TYPES OF MUSCLE*

LACTIC ACID PRODUCTION

In answer to (a) it is not sufficient merely to show that lactic acid is present, because in small concentrations up to 0·02 per cent. it must be considered a common and generally distributed tissue constituent. It is necessary to show that it increases in muscular tissue under those conditions which cause an increase in frog's muscle. The conditions which cause an increase in frog's muscle are: (1) stimulation and fatigue; (2) death, if this occurs at temperatures not too low and leads to *rigor mortis* (Foster and Moyle, 1921); (3) mechanical injury, mincing or chopping; (4) any process that destroys the normal architecture of the muscle without rapidly destroying the enzymes such as heat, if the temperature is not raised too rapidly and too high; chloroform, ether and their like; certain drugs of which caffeine is typical. (2), (3) and (4) all yield about the same amount of lactic acid. A large amount representing complete or nearly complete conversion of glycogen to lactic acid can generally be obtained by (5) suspending the chopped tissue in alkaline phosphate solution.

Reliable values for the lactic acid in resting muscle are difficult to obtain. The best figures for frog's skeletal muscle (Fletcher and Hopkins, 1907; Meyerhof, 1920), skeletal muscle and heart of cat (Hines, Katz and Long, 1925), stomach of tortoise (Evans, 1926), point to about 0·02 per cent. or less, which is roughly the concentration in mammalian blood at rest. Higher resting values for muscles are almost certainly due to deficiencies in technique, owing to the difficulty of removing the muscle from the animal without considerable stimulation (in frog's muscles 100 maximal stimuli produce about 0·1 per cent. lactic acid) and also to the rapidity of production of lactic acid by injury and death.

Lactic acid has been found to increase under the expected conditions in every type of muscle that has been properly examined. Lactic acid has been isolated as a crystalline zinc salt not only from skeletal muscles of frogs and mammals but also from mammalian smooth muscle (Evans, 1925), from the body muscles of cod and haddock, and from the abdominal muscles (striated) of the lobster, *H. vulgaris* (Ritchie, 1927). The lactic acid from lobster was dextrorotatory and indistinguishable from that isolated from cat's muscle. From *Pecten maximus* only a small quantity of crystalline zinc salt can be obtained and that is only partly lactate; the rest is mainly succinate. The material, however, gives the thiophene colour reaction for lactic acid (Boyland, 1928). Henze (1905) obtained only 0·01 per cent. of the zinc salt from *Octopus* muscle. This he considered to be the inactive (fermentation) acid, but his only evidence was the amount of water of crystallization, which is said to be different for the zinc salts of the two acids. In any case the material may have been racemized during isolation. The very small amount of lactic acid obtainable from molluscan muscle cannot by itself be taken as evidence that there is a different mechanism at work. In the genus *Pecten* it is significant that the most active, *P. opercularis*, yields the most lactic acid in fatigue and rigor, the less active *P. maximus* less, while *P. varius*, which lives mostly attached, yields less still.

In other species lactic acid has simply been estimated by oxidation to acetaldehyde and iodometric titration of the aldehyde. The method is not specific for lactic acid and in many cases is not quantitatively exact, but there is no reason to suspect it of serious error. The results of a number of investigations are summarized below. The figures give the normal range of values found in each species (where there are sufficient data) and are expressed as milligrams per gram of tissue. Values obtained by stimulation to fatigue are marked F., in *rigor mortis*, chloroform rigor or similar conditions R., and by treatment with alkaline phosphate solutions to obtain maximum conversion of precursors into lactic acid are marked P. Round numbers only are given because the errors of estimation are considerable and in addition there are large individual variations. Even similar muscles of the same species vary considerably among individuals.

Mammalia

Homo sapiens, skeletal (mixed)   F. about 3   R. 3–5   —   (1) (2)
    The determinations (F.) were made indirectly from the respiratory exchange in exercise to exhaustion, confirmed by estimation of the blood lactic acid.

Felis domestica, skeletal (red)	—	R. 4–7	P. 5–7	(3)
,, heart	—	R. 2–3	P. 3–4	(3)
,, uterus (smooth)	—	R. 1	—	(4)

Lepus cuniculus, skeletal (mixed)   —   R. 4–7   —   (5)
    The red muscles develop less lactic acid than the pale.

Mus norwegicus, skeletal (red)   —   R. 5–6   —   (6)

Aves

Columba livia, pectoral (red)   —   R. 5–8   —   (7)
    ,,   leg (mixed)   —   R. 5–7   —   (7)
    ,,   stomach (smooth)   —   R. 1   —   (7)
Gallus indicus, pectoral (white)   —   R. 8–12   —   (7)
    ,,   leg (mixed)   —   R. 5–7   —   (7)
    ,,   stomach (smooth)   —   R. 1–2   —   (7)
    Among the leg muscles of the fowl the red develop le s lactic acid than the pale.

Reptilia

Testudo sp., skeletal (red)   —   R. 3–5   —   (8)
    ,,   heart   —   R. 1–4   —   (8)
    ,,   stomach (smooth)   F. 1   R. 1   P. 1–2   (4)
Chelhydra serpentina, heart   F. 2   —   —   (9)

Amphibia

Rana esculenta and *temporaria*, skeletal (white)	F. 2–4	R. 4–6	P. 7–13 (10)
R. esculenta, heart	—	R. 1–3	— (8)

    R. esculenta muscle develops rather more lactic acid than *temporaria* and summer frogs (feeding) than winter frogs (starving).

Pisces: Teleostei

Melanogrammus aeglifinus, skeletal	F. 1–2	R. 2–4	P. 2–4 (11) (12)
Gadus callarias, skeletal	—	R. 1–2	— (11) (12)
Urophycis sp., skeletal	—	R. 1	P. 1 (11)
Myoxocephalus sp., skeletal	—	R. 1–6	— (12)

Pisces: Elasmobranchii

Raja clavata, jaw muscle	—	R. 5–7	— (13)
Scyllium canicula, jaw muscle	—	R. 4	— (14)
„ body muscle	—	R. 2	— (14)

Crustacea

Homarus vulgaris, skeletal	F. 1–2	R. 1–3	P. 1–3 (14)
Palinurus vulgaris, skeletal	F. 1–2	R. 2–3	P. 2–4 (14)

    Hearts of *Homarus* and *Palinurus* and skeletal muscle of *Leander serratus* are similar.

Cephalopoda

Eledone cirrosa, body muscle	—	—	P. $<\frac{1}{2}$ (14)
Sepia elegans, body muscle	—	—	P. $<\frac{1}{2}$ (14)

Gastropoda

Buccinum undatum, foot	F. 1	—	P. 1 (14)

Lamellibranchiata

Pecten maximus, quick adductor	F. $\frac{1}{2}$–1	R. $\frac{1}{2}$–1	P. 1–2 (14)
„ slow adductor	—	R. 1	P. 1–2 (14)
„ heart	F. 1	—	P. 5 (14)
P. opercularis, quick adductor	F. 1	R. 1	P. 1–2 (14)
„ slow adductor	F. 1	—	P. 2 (14)
P. varius, quick adductor	F. $\frac{1}{2}$	R. 1	P. 1 (14)

Annelida

Lumbricus sp., body muscle	F. 1	R. 1	P. 3 (14)

Echinoderma

Holothuria nigra, body muscle	F. 1	—	P. 2 (14)

Nematoda

Ascaris megalocephala, body wall	—	R. 8–9	— (15)

    The *Ascaris* figures are obtained in minced tissue under toluene. As the gonad produces even more lactic acid the process of conversion of glycogen to lactic acid is not a specific muscle process in this animal, and provides no relevant information.

References: (1) Hill, Long and Lupton, 1924; (2) von Fürth, 1919; (3) Hines, Katz and Long, 1925; (4) Evans, 1925; (5) Wacker, 1927; (6) Meyerhof and Himwich, 1924; (7) Schmitt-Krahmer, 1927; (8) Arning, 1927; (9) Redfield and Medearis, 1926; (10) Meyerhof, 1920, 1921 and many others; (11) Ritchie, 1927; (12) Macleod and Simpson, 1927; (13) Eggleton and Eggleton, 1928, 2; (14) Boyland, 1928; (15) Fischer, 1924.

CLASSIFICATION OF MUSCLES

The evidence is unfortunately incomplete in many respects. There are no figures for the highly developed muscles of Insects and not enough information about the wing muscles of Birds. The information about lower invertebrate muscle is scanty, but what there is does not suggest that any unexamined group would reveal anything new. Taking the figures as they stand there seem to be four main types of muscle, distinguished by the amount of lactic acid produced under rigor conditions:

These are:

i. Pectoral muscles of Birds; over 6 mg. lactic acid per gram.

ii. Skeletal muscle of terrestrial Vertebrates including Birds' leg muscles; 3–6 mg. lactic acid per gram.

iii. (a) Skeletal muscles of active aquatic animals (Fishes and Crustaceans).

(b) Heart muscle of terrestrial Vertebrates; 1–3 mg. lactic acid per gram.

iv. (a) Body muscles of inactive Invertebrates.

(b) Visceral smooth muscle of Vertebrates; about 1 mg. lactic acid per gram.

The classification, of course, is only rough and provisional and there are bound to be doubtful and transitional cases. But it arranges the tissues for the most part in descending order of activity. The distinction between one muscle and another is functional and not specific. Any given muscle will vary according to its use and according to the age and state of nutrition of the animal. The points arising out of these results that remain to be discussed are first of all the significance of the three different figures: "fatigue", "rigor", and "phosphate", and secondly certain interesting types of muscle.

FATIGUE MAXIMUM

Muscular fatigue is shown by two changes that can be produced by repeated stimulation: (1) diminished *excitability* so that stimuli need to be progressively stronger and stronger to produce a response and finally no response can be obtained by any stimulus; (2) the *relaxation time* lengthens until it may be a hundred times what it was at first. Possibly a third change may take place, a diminution in the tension developed on stimulation, but it is difficult to be certain of this effect, because some fibres may become inexcitable before others and observed changes in tension may be simply changes in the number of fibres excited. This possibility must therefore be left on one side. The two changes in excitability and relaxation time do not necessarily occur *pari passu*. Some muscles become inexcitable before there is any change in relaxation time, some are easily excitable when the relaxation time is greatly lengthened. The difference may be expected to depend on the function of the muscle. A muscle specialized for rhythmic movement is useless as soon as its relaxation time is too long for the maintenance of its rhythm and may be expected to become inexcitable first. A muscle concerned with maintaining tension becomes increasingly efficient the slower its relaxation, particularly if its *refractory period* increases too, because it does not need to be excited so often; each excitation means a fresh output of energy. The contrast is well seen in the two types of *adductor* of the valves of *Pecten*. The large quick muscle produces the flapping movements by which the animal swims, and is specialized for rhythmic movement. By repeated stimulation complete fusion of contractions cannot be obtained and it soon becomes inexcitable, before the relaxation time is increased. On the other hand, the slow muscle which keeps the valves shut gives fused contractions (tetanus) and remains excitable after considerable stimulation when the relaxation time has begun to lengthen (Bayliss and Boyland, 1928). The properties of these muscles will be mentioned again later (p. 81).

It follows that the significance of the stimulation maximum of lactic acid varies in different muscles according to their relative excitability and is not necessarily an index of any property of the contractile mechanism. If we knew the degree of slowing of relaxation produced by a given quantity of lactic acid we should have valid information about that part of the mechanism concerned with relaxation.

RIGOR MAXIMUM

The amount of acid produced in rigor is a more valuable indication. The various processes referred to as rigor, whether spontaneous *post mortem* changes or produced by toluene or chloroform in intact or in chopped muscle or by any other similar treatment, all resemble each other in that there is a breakdown of the normal structure of the tissue whereby the reaction Glycogen → Lactic acid under the influence of the muscle enzymes proceeds as far as it can. There are three possible ways in which the reaction may come to a standstill: (1) exhaustion of the glycogen or other precursor, (2) increasing acidity of the tissue which inhibits the process, (3) destruction of the enzyme or one of the enzymes. There is another possible factor limiting the amount of lactic acid found: (4) partial oxidation or resynthesis of the lactic acid formed. This last possibility has been overlooked but may be of importance if any muscles have an oxidizing mechanism which can work to some extent without free oxygen. However, as there is no evidence this factor (4) must be neglected. As regards (3), there is no reason to suppose any failure of the enzymes in the muscles of the higher animals under the conditions ordinarily used. But in *Pecten* muscle, which contains a large amount of glycogen of which only a little breaks down, something of this sort may occur. On the other hand, the bulk of the glycogen may not be available for conversion to lactic acid under any conditions. (Compare with cephalopod muscle, p. 51.)

This leaves factors (1) and (2) to be considered. In frog's muscles in rigor, as Meyerhof has shown (1921), the whole of the glycogen has not been broken down, but if the reaction be kept neutral with buffer solutions more lactic acid is formed. Phosphates, in addition to their buffering action, have a specific effect and in their presence the whole of the glycogen is converted. If the acid is not neutralized as it is formed the reaction Glycogen → Lactic acid is self inhibiting. The muscle tissue possesses buffer substances but only in limited amount. If therefore a muscle in rigor still contains glycogen or if more lactic acid is formed in the presence of phosphates, then the lactic acid content in rigor is *prima facie* a measure of the buffering capacity of the tissue. This conclusion has been shown to hold good in the comparison of cat's heart and skeletal muscle (Katz, Kerridge and Long, 1925; and Furusawa and Kerridge, 1927, 2) by direct determination of the buffer curves of the tissues, that is to say, by measurement of the change in hydrogen ion concentration produced by a given increment of acid or alkali. The cat's heart gives less acid in rigor not because it contains less glycogen than the skeletal muscle, but because it is less buffered. The heart is an organ with a relatively slow rate of action, it produces moderate tensions only, it has more or less constant work, and has probably a powerful oxidizing system. Normally it may be expected to oxidize the lactic acid as fast as it is produced, keeping up a steady state at a low lactic acid concentration; it has therefore no need for a buffering mechanism to accommodate large quantities of lactic acid. The skeletal muscle, on the other hand, may be called upon to produce lactic acid faster than it can oxidize it and therefore needs the greatest possible buffering capacity.

On this view of the significance of the rigor maximum of lactic acid there appears to be little difference between the skeletal muscles in Group ii. We may say that frog's muscle differs from cat's in being possibly somewhat less buffered and in having a larger glycogen reserve.

The body muscles of Fishes clearly differ greatly from the skeletal muscles of the terrestrial Vertebrates. Leaving out of account the flat fishes and the others that habitually rest on the bottom, the remainder have certain characteristics in common. (1) The animal is nearly always moving and the body muscles performing gently rhythmic contractions, with occasional short bursts of rapid and more violent rhythmic movements. (2) It has no need to maintain a state of *tone* to preserve its posture as land animals have. (3) It possesses an enormous mass of muscular tissue to perform these functions. (4) The tissue has a poor blood supply. (5) The connective tissue between the *myotomes* is fragile and quite different from the tendons attaching mammalian muscle to bone. Generally speaking, the work the muscles have to do more nearly resembles that of heart muscle than that of the limb muscles of land animals. The lactic acid production (see p. 40) is in accordance with this. There are interesting quantitative differences between the three *Gadidae* for which figures are given; the haddock (*Melanogrammus*), cod (*Gadus*) and hake (*Urophycis*). The differences correspond to the general activity and muscular development of the fishes. The hake is a lean fish with little muscle and of sluggish habit; the haddock is stouter and altogether more active; the cod is intermediate but more like the haddock than the hake. It seems that the chief difference is in the available glycogen reserve in muscles. When these fish are in rigor there is only a trace of glycogen left (Macleod and Simpson, 1927) and hardly any more lactic acid can be obtained by treatment with phosphate (Ritchie, 1927). There is no reason to suppose any marked difference in buffering capacity among the muscles of these fishes.

There is unfortunately little information about the muscles of the Elasmobranch Fishes. The jaw muscles (*coraco-mandibularis*) appear from their anatomical arrangement to be suited for the development of high tensions and yield a correspondingly large amount of lactic acid, of the same order as the higher Vertebrates. The body or swimming muscles of the dogfish in this respect, as well as functionally and anatomically, resemble other fish muscles.

Crustacean muscle is not unlike fish muscle, as might be expected. Of the other Invertebrates there is not much to be said except that there is some resemblance to vertebrate smooth muscle. (See p. 40 above for the figures of lactic acid production.)

RED AND WHITE MUSCLE

There remains the distinction between red and pale muscle to consider. (For a review of the subject and literature see Needham, 1926.) It seems fairly obvious that the fact that two muscles are white does not guarantee any further resemblance, nor the fact that one is white and one red any further difference, but however obvious it has sometimes been forgotten. In order to compare red and white muscles we ought to compare muscles that are as far as possible alike in other respects. For instance it would be valuable to have a detailed comparison of the pectoral muscles of the pigeon and the pheasant; it is of far less value to compare the white pectoral muscles of the fowl with its red leg muscles. In the absence of sufficient information about Birds a comparison of the red and pale muscles in a mammalian limb is of interest.

Roberts (1916), who discusses the anatomical distribution and arrangement of red and pale muscles in the rabbit's limbs, points out that the red muscles must be (1) mainly responsible for posture, and (2) in certain cases must act synergetically with pale muscles where the tendon of the pale muscle extends over two joints and one joint needs to be "fixed" for the proper control of movement. The severe mechanical work, however, is done mainly by the pale muscles, which are as a rule larger. It has been known for a long time that mammalian red muscle gives a slower twitch in response to a single stimulation than pale does and gives fused contractions with a slower frequency of stimulation. Fischer (1908) has studied the response of the *soleus* (red) and the *gastrocnemius* (pale) muscles of cats, rats, guinea-pigs, and rabbits. The different animals all gave similar results. A cat's *gastrocnemius* gave in a single twitch an isometric tension of 600 gm. and a total duration of response of about 0·1 second. The *soleus* gave a tension of 200 gm. and a relaxation time of more than 0·2 second. In tetanus, however, the *gastrocnemius* developed a tension of only 700 gm., the *soleus* 950 gm. To obtain complete fusion of contractions of *gastrocnemius* the rate of stimulation must be at least twice that needed for *soleus*. The value of the numerical results is lessened by two considerations. In the first place, the *soleus* is much smaller than the *gastrocnemius* and has (apparently) a simple parallel arrangement of fibres, while the *gastrocnemius* has a complex diagonal arrangement of fibres. This explains why the *soleus* though giving a smaller tension in an isometric twitch can give a greater shortening in an isotonic twitch. In the second place, the difference in response between the two muscles is just as marked in the cat, where both are red and all that can be said is that the *soleus* is rather redder, as it is in the rabbit, where the difference in colour is obvious. It may be that the amount of red pigment or the number of red fibres is only remotely related to other properties of the muscle.

Batelli and Stern (1912) found that chopped red muscle consumed more oxygen and had more active oxidizing enzymes than white. Others have obtained similar results (Needham, 1926). This distinction may be directly related to the amount of red pigment.

If we conceive of mammalian red muscles, or let us say of the redder mammalian muscles, as tissues with powerful oxidizing mechanisms and moderate though long continued work to perform, they will be intermediate between heart muscle and the muscles which have less frequent but more violent duty. Like heart muscle, we might expect them to have a smaller buffering capacity and consequently a lower lactic acid maximum in rigor as Fletcher (1914) first noticed in rabbit's muscles. Wacker (1927) obtained 0·71 per cent. lactic acid as the mean for pale rabbit's muscle in rigor and 0·53 per cent. for red. It may be taken as probable, though there is no direct evidence, that the glycogen reserve was not exhausted and the difference in lactic acid maximum was due to a difference in buffering capacity. Wacker, by titration of boiling water extracts of the muscles, found a greater buffering capacity in pale muscle. The method is not conclusive because it only shows a greater capacity in soluble buffers and leaves nearly all the proteins out of account. The aqueous extract of the pale muscle was more acid in reaction than that of the red, in agreement with the old but excellent observations of Gleiss (1887), so that the question of relative buffering capacity of the two types of muscle is still in doubt.

WING MUSCLES OF BIRDS

In function the leg muscles of Birds are strictly comparable to the limb muscles of Mammals. In accordance with this, Schmitt-Krahmer (1927) finds in rigor a mean value of 0·71 per cent. lactic acid in the pale muscles of the fowl's leg and 0·62 per cent. in the red, similarly 0·54 per cent. in the mixed muscles of the pigeon's leg. The pectoral muscles give higher values in both animals. The pectoral muscles must clearly be treated as a group by themselves; to consider them as similar to the other muscles of the terrestrial Vertebrates confuses the issue. Among different Birds there are interesting functional differences and it is a pity there is not more information about them. For example, taking birds of similar size and all with powers of sustained flight, there is a strong contrast between those with a flapping flight such as the pigeon and kestrel hawk, and soaring birds such as falcons and gulls. The pectoral muscles of the latter are chiefly concerned in maintaining posture, whereas the muscles of the flapping type if active are always in rhythmic movement. Again, among small birds there is a marked contrast between those which continually fly and feed on the wing like swallows and humming birds and those which feed on the ground or in trees and make occasional short flights. Generally speaking the wings of birds seem to be more clearly differentiated and specialized in function than the limbs of mammals. Each mammalian limb has to perform several functions. Though the individual muscles may be differentiated and specialized, their mechanical arrangement is so different (as for instance *gastrocnemius* and *soleus*) that direct comparison is difficult. The pectoral muscles of birds do not differ so much in their anatomical position and structure from one species to another, so that they are more directly comparable and therefore deserve to be investigated. However, this is a digression and the origin of the lactic acid in the muscle has still to be considered.

GLYCOGEN AND LACTIC ACID

The question as to whether glycogen is invariably the source of lactic acid is not easy to answer definitely. The difficulty turns upon the small concentrations that have to be estimated and the inaccuracy of the methods, particularly of glycogen estimation. Meyerhof (1920, 1) found in frog's skeletal muscle that the glycogen could be converted quantitatively to lactic acid and that the changes in other carbohydrates were hardly significant. Expressing the glycogen as $(C_6H_{12}O_6)_n$ the sum of all carbohydrates and lactic acid remained approximately constant in any anaerobic change. This is true also of rat's skeletal muscle (Meyerhof and Himwich, 1924), of skeletal, heart and stomach muscles in the tortoise (*Testudo* sp.) (Boyland, 1928; Evans, 1926), and of the abdominal, skeletal and the heart muscles in the lobster (*Homarus vulgaris*) and the crawfish (*Palinurus vulgaris*) (Boyland, 1928). Some typical experimental results of the authors quoted are given in

Tables 2 and 3. The sums are reasonably constant but there appears to be some tendency for an increase in lower carbohydrate under conditions where glycogen breaks down, particularly when alkaline phosphate is used to produce maximum of glycogenolysis. It should be noted that "lower carbohydrate" simply means the reducing substances found in an alcoholic or aqueous extract after acid hydrolysis and probably includes material that is not carbohydrate.

Table 2

Relations between Carbohydrates and Lactic Acid in Skeletal Muscle

mg. per 100 gm.

		Rat	Frog	Lobster
Low lactic acid				
Lactic acid		60	28	134
Glycogen		710	720	179
Lower carbohydrate		230	188	43
	Sum	1000	936	356
High lactic acid				
Lactic acid		460	288	283
Glycogen		290	387	25
Lower carbohydrate		240	242	70
	Sum	990	917	378

Table 3

Relations between Carbohydrates and Lactic Acid in Tortoise's Muscle

mg. per 100 gm.

		Skeletal	Heart	Stomach
Low lactic acid				
Lactic acid		90	260	20
Glycogen		750	370	160
Lower carbohydrate		160	60	120
	Sum	1000	690	300
High lactic acid				
Lactic acid		790	470	100
Glycogen		20	110	80
Lower carbohydrate		230	150	90
	Sum	1040	730	270

Hines, Katz and Long (1925) did not find sufficient glycogen in cat's heart muscle to account for the lactic acid produced; they did not determine the lower carbohydrate. It appears from their papers that the glycogen and lactic acid were not determined on the same hearts, so that their results are not quite conclusive.

Boyland (1928), however, using pig's heart, confirmed their finding. The extra lactic acid does not come from the "lower carbohydrates", but may come from *inositol*. Inositol is present in resting pig's heart to the extent of 0·3–0·4 per cent., that is, in higher concentration than glycogen, and it diminishes as the lactic acid accumulates. (Compare this with cephalopod muscle, p. 51.)

Bayliss, Müller and Starling (1928), using a mammalian heart-lung preparation, obtained low respiratory quotients which could be raised by administration of glycogen and insulin, but were always less than unity. The experiments do not show what is the R.Q. of the heart's muscular activity or whether it is utilizing fuel other than carbohydrate, because the observed R.Q. is a resultant of (*a*) the heart's resting metabolism, and more important, (*b*) the metabolism of the lungs, as well as (*c*) the heart's active metabolism. Until (*a*) and (*b*) are known no conclusion as to (*c*) is possible. Thus although it seems that the mammalian heart *can* obtain lactic acid from a source other than glycogen, it is not clear that it does so normally, much less that the other source is not carbohydrate. If inositol is a source of lactic acid it is still a substance very closely allied to carbohydrate and gives the same R.Q. on combustion. At the most the mammalian heart is under suspicion of having a somewhat abnormal metabolism.

Turning to invertebrate muscle the work of Boyland (1928) shows that there are probably few anomalous types of muscle. Crustacean muscle both skeletal and heart behaves normally, the lactic acid all comes from glycogen. The glycogen content of the muscle, however, is liable to fluctuations depending on the reproductive and growth cycles of the animal. The formation of a new exoskeleton at the moult calls for carbohydrate for laying down of fresh *chitin* and the muscles seem to store glycogen for this purpose.

Among the Molluscs there are some peculiarities. The large

adductor of *Pecten* has a variable and often high glycogen content, up to 2 or 3 per cent., only a very small portion of which can be converted to lactic acid. The bulk of the glycogen seems to be general storage material and no part of the muscular mechanism. In specimens rich in glycogen some of it passes readily into a form in which it escapes estimation as either "glycogen" or "lower carbohydrate"; probably a form of dextrin. The small slow adductor muscle of *Pecten* contains less glycogen but yields more lactic acid and behaves in quite a normal way, though the glycogen is not generally completely converted to lactic acid even with alkaline phosphate treatment. The heart muscle is similar to the slow adductor. The muscle of the foot of the whelk, *Buccinum undatum*, resembles the large adductor of *Pecten*.

Cephalopod muscles differ from other molluscan muscles or indeed any muscles. Henze (1905) failed to find glycogen in *Octopus* muscle, but later Starkenstein and Henze (1912) isolated it from the muscles of another Cephalopod, *Eledone*. Glycogen is undoubtedly present but in small amounts. The animals are extremely active, do not bear handling at all well, and the muscles die quickly. Probably the muscles when they are analysed are always in a state of "fatigue". It is possible too that some of the muscle enzymes are unstable. The results of an experiment by Boyland (1928) on *Sepia elegans* are worth quoting for their suggestiveness, though too much stress should not be laid on the figures as the concentrations of the various constituents are very small and the analytical methods not very accurate, particularly that for inositol. The whole body muscle of the animal was used. The results are given in Table 4.

Table 4

mg. per 100 gm. of Tissue

Treatment of tissue	Lactic acid	Gly-cogen	L. carbo-hydrate	Inositol	Total
"Rest"	15	24	34	70	143
CHCl₃ rigor	21	32	44	50	147
Alk. phosphate	29	34	46	40	149

Similar amounts of lactic acid and glycogen were found in the arm muscle of *Eledone cirrosa* but rather more glycogen (0·11 per cent. in "fatigue") in the muscle of the head. Altogether cephalopod muscle appears the most peculiar of all muscles so far examined. Even if we assume that under experimental conditions the enzymes are not working properly the very small lactic acid concentrations are remarkable in such an active muscle, but are not such as to rule out the probability of lactic acid playing its usual part in the muscle processes. There is of course the possibility that lactic acid does not accumulate because it is removed by some kind of partial recovery process.

A few experiments on the body muscles of earthworms (*Lumbricus* sp.) and on the longitudinal body muscles of *Holothuria niger* revealed nothing peculiar in the carbohydrate breakdown (Boyland, 1928).

The general conclusion is that within the two great groups of Vertebrates and Arthropods in all cases so far examined glycogen is present in muscle and when glycogen breaks down lactic acid appears. If the disappearance of the one and the appearance of the other do not correspond quantitatively, the difference is probably but not certainly to be attributed to intermediates between glycogen and lactic acid. Outside these groups the evidence is scanty and less clear. We are dealing with animals that are generally sluggish (except the Cephalopods) and whose tissues are less specialized. Allowing for this there seems no definite reason for suspecting a fundamentally different chemical mechanism.

It must be remembered, however, that it is only in frog's skeletal muscle and in snapping turtle's heart muscle (Redfield and Medearis, 1926) that there is direct experimental evidence to show that lactic acid is the chief causal factor in muscular contraction: that is to say, that the tension developed varies as the lactic acid produced. In all other cases there is direct evidence as to chemical changes only and the analogy of these changes with the changes in frog's muscle is the reason for assuming a similar mechanism. But there is the additional evidence that the ratio of heat production to tension in whatever muscles measured has been found closely similar (Hill, 1926). This suggests a similarity of chemical mechanism.

OXIDATIVE PROCESS

The question of the oxidative removal of lactic acid and the synthesis of glycogen from it may be attacked both from the calorimetric and chemical point of view; and chemical evidence can be obtained directly by study of isolated tissues and indirectly by a study of the gaseous metabolisms of the whole organism. Taking the three methods in order.

Hartree and Liljestrand (1926) have found that the heat production of *oxidative recovery* in tortoise's skeletal muscle is the same as in the frog, namely 1·5 times the *initial* heat. Tortoise's muscle differs from frog's muscle in its time scale but not apparently in any other respect.

Evans (1925) has found an excess oxygen consumption in tortoise's stomach muscle during oxidative recovery. The fraction of lactic acid oxidized is 1/3·35 of that produced, not appreciably different from the figures for frog's muscle.

In man the oxygen intake and the carbon dioxide output after hard exercise show an oxidative recovery process closely resembling that in frog's muscle. If the effort made exceeds a certain limit the rate of oxygen intake cannot be increased and the effort is necessarily short lived. This means that the oxygen requirements of the muscles are more than the maximum output of heart and lungs can satisfy. Under these conditions lactic acid accumulates in the blood, presumably as the result of its accumulation in tissues, and there is an excess carbon dioxide output due to displacement from bicarbonates in the blood. After exercise is over the lactic acid in the blood disappears, and there is a deficiency of CO_2 output, as CO_2 displaces lactic acid in the blood. In the meantime there is an excess oxygen consumption over the resting value. This excess oxygen, the *oxygen debt*, is that needed for the oxidative recovery. If we assume that the efficiency of the oxidative recovery process is the same in man as in the frog we can use the values obtained from oxygen debt to calculate the amount of lactic acid that has accumulated in the body during exercise. An approximate estimate of the concentration in the active muscles can then be made. The figure given on p. 39 is obtained in this way. The general conclusion is that the human muscle closely resembles the muscle of the other higher vertebrates.

Davis and Slater (1927) have found that the cockroach (*Periplaneta orientalis*), which survives for a period in nitrogen, is accumulating an oxygen debt in the same way and consumes excess oxygen afterwards when restored to air to compensate exactly for the anaerobic period.

In all probability all the cases of alleged anaerobiosis in the animal kingdom are to be explained as due to the fact that animals with a slow rate of metabolism can accumulate an oxygen debt for many hours. There does not seem to be any evidence that will stand criticism for a genuine permanent anaerobiosis among animals (see Fischer, 1924, and Slater, 1925). Collip (1920, 1921) and Berkeley (1921) have studied the metabolism of the clam (*Mya arenaria*) and of other molluscs. *Mya* will survive anaerobic conditions for 24 hours or more; during this time glycogen disappears, the combined CO_2 in the body fluids increases without appreciable change of pH, and CO_2 is excreted. On recovery in oxygen excess oxygen is consumed. There is no difference qualitatively between the behaviour of *Mya* and that of man or isolated frog's muscle. Collip considers that the calcium carbonate of the liver and possibly of the shell, too, plays the part of a large "alkali reserve". He does not discuss lactic acid metabolism but it is obvious that a conversion of calcium carbonate into calcium lactate and soluble calcium bicarbonate would explain all the results.

Davis and Slater (1928, 1) have recently compared the lactic acid produced in cockroaches (*Periplaneta orientalis*) during a period of anaerobiosis with the excess oxygen consumption afterwards. The oxygen consumption is greatly in excess of that needed in the ordinary recovery process of muscle, in fact it agrees with that required for the complete combustion of the lactic acid. In that case these insects may have no recovery process in the sense that they do not reconvert lactic acid to glycogen. These experiments, it should be said, are really determinations of the resting metabolism of the animals not of the metabolism of muscular activity, because they are inactive under anaerobic conditions.

On the other hand a normal oxidative recovery process was found in the earthworm (Davis and Slater, 1928, 2).

USE OF FAT BY MUSCLES

The last question to be considered is whether the muscles can utilize any other source of energy besides carbohydrates, fat for instance. Experiments on isolated muscle have so far failed to produce evidence that fats can be utilized (Winfield, 1915). Meyerhof and Himwich (1924) found that in rats fed on a fatty diet, the muscle glycogen was diminished and the amount of lactic acid liberated in rigor was diminished in proportion. It might happen that when carbohydrate was scarce the glycogen-lactic acid mechanism in muscle would still be used for contraction but that carbohydrate would be conserved by oxidizing fat to supply the energy for resynthesis. In ordinary muscle carbo-hydrate is both part of the machine and the fuel, but when carbo-hydrate supply was lacking it might be used as machine only and the fires be stoked with fat. On the other hand the fat if used for muscular work might be first converted into carbohydrates. The problem has been studied in man by many workers (Atwater and Benedict, 1903; Zuntz, 1900; Benedict and Cathcart, 1913; Campbell, Douglas and Hobson, 1921; Krogh and Lindhard, 1920; Himwich, Loebel and Barr, 1924; Furusawa, 1925). Krogh and Lindhard, by measuring the cost of maintaining a given moderate intensity of work in trained subjects and by accurately measuring the respiratory quotients of these subjects on different diets, found that the total energy used for unit work performed, varies inversely with the respiratory quotient. That is to say, carbohydrate only is used if it is available and then the efficiency of the oxidative process is maximal. The efficiency falls lower as a larger proportion of fat is utilized. They concluded that fat was used only after conversion to carbohydrate, a process involving a loss of 10 per cent. of the energy of the fat.

Furusawa studied the excess metabolism during and after violent exercise on various diets. On a normal diet the R.Q. at rest is fairly low (0·8–0·9) in exercise of short duration the R.Q. of the excess metabolism approximates very closely to 1, indicating the use of carbohydrate only. In exercise of longer duration it tends to fall after a time. On a fatty diet the resting R.Q. can be brought down nearly to the value for pure fat metabolism (0·7).

For exercise of short duration, 0·3–2·0 minutes, the R.Q. rises to 1 as on an ordinary diet. If the exercise is prolonged to 5–10 minutes it falls to a value about 0·92. These results confirm Krogh and Lindhard in the view that carbohydrate is always used if available, but that fat may be called upon when carbohydrate reserves are exhausted. The further conclusion that fat is first converted to carbohydrate seems probable but it is not established. It is suggestive that the isolated liver appears to be able to form sugar from some source that is not protein, carbohydrate or lactic acid and therefore is probably fat (Burn and Marks, 1926).

Experiments on man indicate that there is a small increase of nitrogen metabolism accompanying muscular exercise (Cathcart, 1925). Whether this is actual muscle metabolism or a secondary effect of it is impossible to say. Relative to the amount of carbohydrate metabolism the amount of hydrogen metabolism is insignificant. It may perhaps be connected with the small ammonia production which has been observed in isolated muscles (see p. 21).

REFERENCES

REVIEWS AND MONOGRAPHS

CATHCART (1925). *Physiol. Rev.* **5**, 225.
EVANS, C. L. (1926). *Recent Advances in Physiology.* 2nd ed. London.
FULTON, J. F. (1926). *Muscular Contraction.* Philadelphia and London.
HILL, A. V. (1916). *Ergeb. d. Physiol.* **15**, 340.
— (1922). *Physiol. Rev.* **2**, 310.
— (1925). *Aspects of Biochemistry.* London. p. 253.
— (1926). *Muscular Activity.* Baltimore.
HILL, A. V. and MEYERHOF, O. (1923). *Ergeb. d. Physiol.* **22**, 299.
MACLEOD, J. J. R. (1926). *Carbohydrate Metabolism and Insulin.* London.
MEYERHOF, O. (1924). *Chemical Dynamics of Life Phenomena.* Philadelphia and London.
NEEDHAM, D. M. (1926). *Physiol. Rev.* **6**, 1.

PAPERS

ARBEITER (1921). *Arch. Néerland.* **5**, 185.
ARNING (1927). *J. Physiol.* **63**, 107.
ATWATER and BENEDICT (1903). *U.S. Dep. Agric. Bull.* No. 136.
BATELLI and STERN (1912). *Bioch. Zeit.* **46**, 317.
BAYLISS, L. E. and BOYLAND (1928). Unpublished.
BAYLISS, L. E., MÜLLER and STARLING (1928). *J. Physiol.* **65**, 33.
BENEDICT and CATHCART (1913). *Pub. Carnegie Inst. Wash.* No. 187.

BERKELEY (1912). *J. Biol. Chem.* 46, 579.
BEST, HOET and MARKS (1926). *Proc. Roy. Soc.* B, 100, 32.
BISHOP and GILSON (1927). *Am. J. Physiol.* 82, 478.
BOYLAND (1928). *Biochem. J.* 22, 362.
BURN and DALE (1924). *J. Physiol.* 59, 164.
BURN and MARKS (1926). *Ibid.* 61, 497.
CAMPBELL, DOUGLAS and HOBSON (1921). *Phil. Trans.* B, 210, 1.
COLLIP (1920). *J. Biol. Chem.* 45, 23.
— (1921). *Ibid.* 49, 297.
DAVIS and SLATER (1927). *Biochem. J.* 20, 1167.
— — (1928, 1). *Ibid.* 22, 331.
— — (1928, 2). *Ibid.* 22, 338.
EGGLETON and EGGLETON (1927). *Biochem. J.* 21, 190.
— — (1928, 1). *J. Physiol.* 65, 15.
— — (1928, 2). Private communication.
EINTHOVEN and HUGENHOLTZ (1921). *Arch. Néerland.* 5, 174.
EMBDEN (1927). *Klin. Woch.* 6, 628.
EMBDEN and ZIMMERMANN (1924). *Zeit. Physiol. Chem.* 141, 225.
— — (1927). *Ibid.* 167, 114, 137.
EVANS, C. L. (1925). *Biochem. J.* 19, 1115.
— (1926). *Physiol. Rev.* 6, 358.
FENN (1924). *J. Physiol.* 58, 175.
FISCHER, A. (1924). *Biochem. Zeit.* 144, 224.
FISCHER, H. (1908). *Pflüg. Arch.* 125, 541.
FISKE and SUBBARROW (1927). *Science,* 65, 401.
FLETCHER, W. M. (1898). *J. Physiol.* 23, 10.
— (1914). *Ibid.* 47, 361.
FLETCHER and HOPKINS (1907). *Ibid.* 35, 247.
— — (1917). Croonian Lecture, *Proc. Roy. Soc.* B, 89, 444.
FOSTER and MOYLE (1921). *Biochem. J.* 15, 672.
VON FÜRTH (1919). *Ergeb. d. Physiol.* 17, 387.
FURUSAWA (1925). *Proc. Roy. Soc.* B, 98, 65.
FURUSAWA and HARTREE (1926). *J. Physiol.* 62, 203.
FURUSAWA and KERRIDGE (1927, 1). *J. Mar. Biol. Ass.* 14, 657.
— — (1927, 2). *J. Physiol.* 63, 33.
GERARD (1927). *Ibid.* 280; *Am. J. Physiol.* 82, 381.
GLEISS (1887). *Pflüg. Arch.* 41, 69.
HARTREE and HILL (1921, 1). *J. Physiol.* 55, 133.
— — (1921, 2). *Ibid.* 389.
— — (1922, 1). *Ibid.* 56, 294.
— — (1922, 2). *Ibid.* 367.
— — (1923, 1). *Ibid.* 58, 127.
— — (1923, 2). *Ibid.* 441.
— — (1923, 3). *Ibid.* 470.
HARTREE and LILGESTRAND (1926). *Ibid.* 62, 93.
HENZE (1905). *Zeit. Physiol. Chem.* 43, 477.
HILL (1912). *J. Physiol.* 44, 466.

HILL (1913). *J. Physiol.* **46**, 28.
— (1914). *Ibid.* **47**, 305.
— (1925, 1). *Ibid.* **60**, 237.
— (1925, 2). *Proc. Roy. Soc.* B, **98**, 506.
— (1926). *Ibid.* **100**, 87.
HILL, LONG and LUPTON (1924). *Proc. Roy. Soc.* B, **97**, 84.
HIMWICH, LOEBEL and BARR (1924). *J. Biol. Chem.* **59**, 265.
HINES, KATZ and LONG (1925). *Proc. Roy. Soc.* B, **99**, 8, 20.
KATZ, KERRIDGE and LONG (1925). *Ibid.* 26.
KROGH and LINDHARD (1920). *Biochem. J.* **14**, 290.
MACLEOD and SIMPSON (1927). *Contrib. Canad. Biol.* **3**, 439.
MACSWINEY and MUCKLOW (1922). *J. Physiol.* **56**, 397.
MEYERHOF (1919). *Pflüg. Arch.* **175**, 20, 88.
— (1920, 1). *Ibid.* **182**, 232, 284.
— (1920, 2). *Ibid.* **185**, 11.
— (1921). *Ibid.* **191**, 128; **195**, 22.
— (1922). *Biochem. Zeit.* **129**, 594.
— (1924). *Pflüg. Arch.* **204**, 295.
— (1927). *Brit. Med. Journ.* Nov. 12, p. 859.
MEYERHOF and HIMWICH (1924). *Pflüg. Arch.* **205**, 415.
MEYERHOF, LOHMANN and MEIER (1925). *Biochem. Zeit.* **157**, 459.
MEYERHOF and MEIER (1924, 1). *Ibid.* **150**, 233.
— — (1924, 2). *Pflüg. Arch.* **204**, 448.
MEYERHOF and SCHULZ (1927). *Ibid.* **217**, 547.
MINES (1912). *J. Physiol.* **46**, 188.
OLMSTEAD and HARVEY (1927). *Am. J. Physiol.* **80**, 643.
REDFIELD and MEDEARIS (1926). *Ibid.* **77**, 662.
RITCHIE (1927). *Brit. J. Exp. Biol.* **4**, 327.
ROBERTS (1916). *Brain,* **39**, 335.
SCHMITT-KRAHMER (1927). *Biochem. Zeit.* **180**, 272.
SLATER (1924). *Biochem. J.* **18**, 621.
— (1925). *Ibid.* **19**, 604.
SOSKIN (1927). *Am. J. Physiol.* **81**, 382.
STARKENSTEIN and HENZE (1912). *Zeit. Physiol. Chem.* **82**, 417.
SURANYI (1926). *Pflüg. Arch.* **214**, 228.
WACKER (1927). *Biochem. Zeit.* **184**, 192.
WINFIELD (1915). *J. Physiol.* **49**, 171.
ZUNTZ (1900). *Pflüg. Arch.* **83**, 557.

Chapter III

MECHANICS AND THEORY OF MUSCULAR CONTRACTION

§ I. *MECHANICS OF MUSCLE*

The chemical and heat changes that occur in muscle may be of the nature of general cell processes. What is specific to muscle is the ability to convert the energy of the chemical breakdown into external work by shortening and developing tension. Consequently a study of the mechanical changes is the only means of revealing the intimate mechanism of the muscle considered as such and as distinct from any other living cell. Though there is no branch of muscle physiology which can be discussed intelligently without reference to the work of A. V. Hill and his colleagues, this particular branch they have made almost entirely their own.

Before any further discussion of the processes in muscle which follow upon excitation it is advisable to say something of the excitation process itself as far as it concerns muscle excluding, that is, the processes in nerve and in the neuro-muscular junction. As regards the relation between the stimulus and the response of the muscle we may take for granted the all or none behaviour of the fibres of skeletal muscle (Lucas, 1909; Pratt and Eisenberger, 1919; Adrian, 1922; Porter and Hart, 1923) and of the vertebrate heart, but it is as well to see what is implied in this. As was pointed out by Adrian (1914) if an excitable tissue has (1) a definite *threshold value* for stimulation below which no significant change occurs, and (2) after excitation an absolute *refractory period* such that no stimulus whatever produces any change, then it follows that a graded stimulus cannot produce a graded response in a single element. That is to say, once the threshold has been reached the excitation process will begin and spread throughout the unit of tissue quite independently of any subsequent stimulation and this is what the experimental evidence confirms. But of course the excitation process is not momentary and does not occur at once over the whole unit. It runs a definite course, increases to a maximum and dies away; at the same time it spreads from the seat of initial excitation at a finite speed. The quantity of energy liberated and the time course of the changes will depend upon the internal state of the tissue and may vary with internal events subsequent to the first moment of excitation. The response of the muscle fibre is all or none with respect to external events but not with respect to internal events.

FREE ENERGY OF EXCITED MUSCLE

If it be assumed that on stimulation a muscle fibre becomes a new elastic body with an increased potential energy varying with its length, then it is possible to construct an "indicator diagram" of the muscle by plotting tension against extension for resting and for excited muscle. The area enclosed by the curves then measures the potential energy set free by excitation.

Using frog's *sartorius* muscle, which is especially suitable for work of this kind as its fibres run parallel to its length, it is found that the figures for the energy obtained by this method agree with the mechanical equivalent of the heat produced during activity in isometric contraction. This suggests that isometrically the whole of the chemical potential energy is converted to elastic potential energy; but the case is not so simple. Fenn (1924) has shown that, if the muscle is allowed to shorten and perform work, more heat is produced than under isometric conditions. The excited muscle cannot therefore be treated simply as a new elastic body of constant properties whatever its length. The amount of chemical breakdown depends not only on the initial conditions at the moment of stimulation but varies with subsequent events. The "indicator diagram" in its simple form may be misleading and in any case cannot be quantitatively exact (Hill, 1925, 1 and 1926, 1).

The relation between heat production and the events after stimulation has been further studied by Azuma (1924), Hartree (1925) and Wyman (1926). The general conclusion is that more heat is produced if the muscle shortens during contraction or lengthens during relaxation; less heat is produced if it lengthens (is pulled out) in contraction or shortens in relaxation. The effects might be simply due to the special thermo-elastic properties of resting and excited muscle (see Chapter 1, p. 17) but is more probably to be explained on the lines suggested by Wyman; namely, that during relaxation part of the energy of the excited muscle (about one-third) is not dissipated as heat but can be absorbed in some type of recovery process. These phenomena, first observed by Fenn, show also that the "all or none law" does not apply to muscle without qualification, as, given constant initial conditions, the response of a single fibre is not fixed but varies according to what happens after the stimulus has taken effect.

REALIZABLE WORK

The *realizable work* varies with the conditions of loading. The highest values are obtained by opposing to the muscular tension at each moment a load which the muscle can just move. This can be arranged by opposing to the muscle the inertia of a mass, as in the flywheel device for human muscles devised by Hill (1922). The muscle does not raise a weight against gravity but sets a heavy wheel spinning. The reaction of the wheel is always just equal to the force exerted. The maximum speed of revolution of the wheel gives the work done. By varying the equivalent mass of the wheel the rate of shortening can be varied. If the pull is exerted on a pulley of large diameter the wheel "feels light" and the contraction is rapid; with a pulley of smaller diameter it "feels heavier" and the contraction is slower.

By this means it can be shown that there is an optimum speed of contraction for any type of muscle. In a single muscle twitch, produced by an instantaneous stimulus, energy is expended in developing tension or in shortening. If the muscle is stimulated by a series of stimuli, so as to maintain the tension or keep shortened, energy is expended in this process also. The energy needed to maintain tension is less than that needed to develop it, but maintaining tension does not in itself produce external work. Consequently, if the process of contraction is prolonged for more than a certain time the efficiency of the process in producing external work falls off. On the other hand, the contractile process is accompanied by internal changes in the viscous substance of the muscle and energy is spent irreversibly in overcoming the frictional resistance. The quicker the movement the greater the loss from this cause. Consequently, with increasing speed the efficiency of movement falls off very quickly.

Curves obtained for human muscles (the flexors of the arm) show considerable individual variations, but are all of the same type. If to any individual we allot a "strength constant" λ, which will depend upon his size and muscular development, then his curve will fit a "standard" curve in the sense that he can impart to mass $M\lambda$ the same velocity as the "standard" individual can to mass M. Taking the mean curve for twenty-two men as standard, Hill (1922) found that the mean "strength constant" of eight women was about 0·5.

The relation between mechanical efficiency of human muscle and speed of movement shows a distinctly blunt peak, so that there is a considerable range of speed over which efficiency is about maximal. In voluntary movement the speed of contraction is probably normally adjusted to fall within this region; if this is so it would account for the relatively close agreement for the mechanical efficiency of human movement found by different observers using different methods. Lupton (1923) has pointed out that if the contraction of human muscle is completed in as short a time as 0·26 sec. the external work is nil, yet the muscles of small mammals and birds contract much faster than this and the wing muscles of insects immensely faster. Human muscles are relatively slow. Among similar animals, speed of movement generally diminishes with increase of size, a relation that is to be expected from the Principle of Dynamical Similarity. Somewhat similar curves relating work and load or work and speed have been obtained with frog's muscle (Doi, 1920; Gasser and Hill, 1924).

HEAT AND TENSION

The question of the relationship between the energy required to develop and to maintain tension has also been investigated in frog's muscle by Hartree and Hill (1922) by studying the variation of the heat production with duration of stimulation.

The matter can be expressed by the equation

$$H/Tl = A + Bt,$$

where H is heat production, T tension, l length of muscle and t the duration of stimulation. A and B are constants. For a momentary stimulus, that is a single twitch, $H/Tl = A$. The mean value of this ratio, taking for convenience the reciprocal Tl/H, is 6. The extreme variation with frog's sartorius is between 4 and 8. The value of A is independent of temperature and lies within the same range for many muscles; namely, leg muscles of frog and tortoise, jaw muscles of *Scyllium canicula*, the dogfish, and the retractor muscle of the foot of *Mytilus* (Hill, 1926, 1). Though the value of the ratio H/Tl is independent of temperature for a single twitch, the actual amount of heat is greater the lower the temperature, presumably because the "valve" controlling the energy output works more slowly.

The value of B increases rapidly as the temperature increases. Since tension is maintained by a fusion of twitches and the twitches occupy less time the higher the temperature, it follows that at a higher temperature more twitches are needed in unit time. We might expect therefore that in slow muscles of the type that can maintain a tension for a long time with small expenditure of energy the value of B will be smaller than in muscle of a rapid type. B in fact is a measure of the time scale of the muscular process in different types and under different conditions.

A is a function of the contractile mechanism as such and is (more or less) independent of variations in type of muscle and in condition, provided the mechanism is intact and in working order. This proviso is necessary because, as Mashimo (1924) showed for frog's muscle and Hill (1926, 2) for that of the dogfish, the ratio H/Tl increases with fatigue and with the length of time the muscle has been exercised, that is, with a change for the worse in the internal state of the muscle. The chief respect in which frog's muscle differs from dogfish's muscle, and many other muscles, is that it deteriorates sufficiently slowly for good measurements to be made on it. Under conditions in which the muscle deteriorates heat and tension diminish together showing that they are intimately connected. The ratio H/Tl gets larger, that is, the efficiency diminishes, but as long as heat is liberated some tension is developed, though it may be very little.

VISCOUS-ELASTIC PROPERTIES OF MUSCLE

It has been mentioned that with increasing speed of contraction the efficiency of muscular contraction diminishes and that this effect is to be attributed to viscosity. Muscle, like many other materials, is both viscous and elastic. These properties have been studied by Gasser and Hill (1924), Hill (1926, 3) and Levin and Wyman (1927) chiefly by the method of stretching or releasing a muscle at various times after excitation by a single stimulus or during continued stimulation.

Rapid stretching of a frog's sartorius muscle just before and at the moment of stimulation produces no special effect beyond lengthening it. The mere passive lengthening of a muscle, it must be mentioned, results in a longer duration of twitch and some increase in tension. But if the stretching takes place a few thousandths of a second ($5-15\ \sigma$) after stimulation quite a different type of effect is produced, namely, a large increase in tension. In other words the excited muscle is a much less extensible body than the unexcited and the new elastic state is produced very quickly and disappears very quickly. The production of tension as ordinarily measured lags behind owing to the viscosity of the muscle substance so that by the time the maximum tension is attained the fundamental change is rapidly dying away. The time course of the "fundamental" change as revealed by the change in extensibility must correspond very nearly to that of the electrical response.

If the muscle is released suddenly during a prolonged contraction the tension drops momentarily to a very low value and then gradually recovers to that proper to the contracted muscle at the new length. The resting muscle passively stretched and released comes to equilibrium much more quickly. That is to say, the excited muscle is both more viscous and less extensible than the resting.

The experiments on quick stretch and release have been repeated by Hill (1926, 3) on the longitudinal body muscles of *Holothuria nigra*. These are unstriated muscles which give a very slow response on stimulation. Allowing for the difference in time scale the phenomena are not notably different from those of frog's muscle. He concludes, " ...we should need to look for the cause of the vast differences existing between different contractile tissues, not in any mysterious difference in their mechanisms, but simply in the specific structural chemistry of the material which goes to form their protoplasm". Hill's suggestion that the mechanics of muscular contractions are governed by viscosity as well as by elasticity has been further developed by Levin and Wyman (1927). In a mathematical and experimental treatment of the subject they show that the facts can be accounted for on the assumption that the muscle is a complex system consisting of a free-elastic part comparable to an undamped spring and a viscous-elastic part comparable to a highly damped spring. The viscous-elastic part is probably the contractile mechanism. The evidence, obtained from experiments on several different types of muscle, is briefly as follows.

From a stimulated muscle stretched and released at different speeds between two fixed lengths a set of tension-length curves can be obtained. On release the tension falls lower the quicker the release and similarly rises higher the quicker the stretch. The tensions for release are always lower and those for stretch always higher than the tension at the smaller and greater lengths respectively when the lengths have been reached infinitely slowly. Treating the curves as "indicator diagrams" the work done by or on the muscle at different speeds can be calculated. Typical tension-length and work-speed curves copied from Levin and Wyman's paper are shown in Figs. 1 and 2. In a simple viscous-elastic system both kinds of curve would be linear, as can be shown theoretically and experimentally on a model. A complex system that contains free-elastic as well as viscous-elastic elements, such as an undamped spring pulling on a damped one, produces exponential tension-length curves and an S-shaped work-speed curve, similar to those given by muscles. This again can be shown theoretically and experimentally.

A further resemblance between the theoretical requirements of a complex free-elastic and viscous-elastic system and the behaviour of muscle is that the tension-length curves for different speeds all start at the same angle. This angle should be independent of viscosity and depend only on the constants of the free-elastic

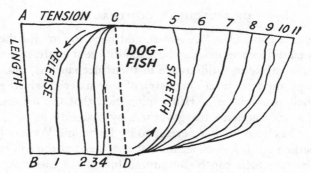

Fig. 1. *Tension-length* curves for dogfish's muscle (Levin and Wyman, 1927).

The curves are drawn from the experimental tension-length curves and super-imposed so as to bring the axes of ordinates and abscissae into coincidence. Each curve represents a stretch or release at constant speed. The broken line CD represents the ideal curve for zero speed. The shorter broken line is curve 4 corrected for fatigue which affected the release at the slowest speed. The line AB corresponding to stretch or release at zero tension is obtained by joining the points of zero tension at the two lengths between which movement takes place. The numbers indicate the order in which the observations were made. The corresponding speeds, in arbitrary units, are as follows: (1) 6·15, (2) 0·70, (3) 0·27, (4) 0·13, (5) 0·17, (6) 0·25, (7) 0·49, (8) 1·42, (9) 3·47, (10) 3·69, (11) 6·71.

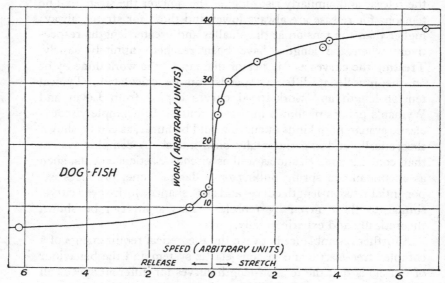

Fig. 2. *Work-speed* curve for dogfish's muscle plotted from results shown in Fig 1 (Levin and Wyman, 1927).

system. It might be expected therefore to be independent of temperature. This is so between 7·5° and 21·5° C.

Gasser and Hill (1924) found that quick release of stimulated frog's muscle produced a drop in tension followed by an increase along a curve corresponding to that of the initial rise of tension after stimulation, that is if we neglect, as is legitimate, the initial inflected portion of the tension-time curve while the new excited state is developing and consider only the later portion which is logarithmic in both cases. If the extent of release is less than 15 per cent. of the resting length of the muscle, the tension does not fall to zero. In a simple viscous-elastic system the tension would drop to zero after any instantaneous release however small, because the system would take time to get into motion. It is only an undamped-elastic element that can preserve a residual tension after a quick release. Gasser and Hill's experiment can be taken in accordance with this theory to show that when a frog's sartorius muscle contracting isometrically attains its maximum tension, the free-elastic elements are stretched by an amount equal to 15 per cent. of the resting length of the muscle.

The change on stimulation must be in the viscous-elastic part of the system, because, as Levin and Wyman argue, if the free-elastic system only were concerned there would be a sudden development of tension on stimulation followed by a drop to the equilibrium value, instead of a steady rise as actually occurs.

Repeated releases at constant speed of a muscle stimulated to fatigue showed an increase in viscosity parallel to fatigue development.

Most of the experiments were done on the jaw muscles of the dogfish, *Scyllium canicula*. They are striated and fairly quick moving. Experiments were also done on the longitudinal body muscles of *Holothuria nigra* and the lantern muscles of *Echinus*. They are unstriated and slow moving. These muscles differ widely in speed but are probably reasonably homogeneous in structure and are adapted to considerable changes in length. They are therefore well suited for stretch and release experiments. Except for an explicable discrepancy to be discussed shortly these muscles all gave qualitatively similar curves agreeing with theory. Other muscles, the claw muscles of the spider crab, *Maia*, and the biceps cruris of the tortoise were less satisfactory, but for causes that can almost certainly be accounted for. However, all the tissues give exponential tension-length curves with constant initial angles and an S-shaped work-speed curve. The curves do not correspond to those of a simple viscous-elastic system but do correspond reasonably closely to those of the complex system. The discrepancies are matters of detail, but need to be discussed.

In the first place, the theoretical treatment assumes no distinction but one of sign between tension and compression and the illustrative model of damped and undamped springs is equally capable of compression or tension. The muscle however is not; if released too quickly it will buckle. If we make the probable assumption that the fibres constituting a muscle vary appreciably one from another in mechanical properties, we can see that on release some of the slower fibres may lag behind and make no contribution to the tension, though the muscles as a whole will not buckle. Consequently at the faster speeds fewer fibres will be contributing to the total effect and the tensions will be smaller than they should be. The theoretical work-speed curve is symmetrical for stretch and release on either side of the point of infinitely slow change of length. The work-speed curve for dogfish muscle, as Fig. 2 shows, deviates from the theoretical curve in the direction of giving tensions that are too small at higher speeds of release. Holothurian and *Echinus* muscles are similar.

With spider crab's and tortoise's muscle a different discrepancy is found, the tensions on stretch are too small. Levin and Wyman explain this as due to slip or rupture of some structure; but as they found that normal tensions could be obtained in later stimulations, it follows that this slip must be reversible. It must be explained that the supposed slip is not analogous to the rupture or yield of a stretched solid, which is not reversible. The excited state of the muscle is unstable, it is tending to pass back into the unexcited state, so that tension is maintained only by continual renewal of a structure which is continually breaking down. It is possible to imagine that in such a system there might be some sort of slip of the excited structure leading to a lowering of tension at the time, but not to any permanent change once excitation is over and the muscle is back in its resting condition.

§ II. *THEORIES OF MUSCULAR CONTRACTION*

Theories of muscular contraction have generally assumed that the essential process was a change affecting the relative position or character of internal elements of the muscle fibre. Dismissing a few older theories which have no basis in fact, and have apparently no support at the present time, possible theories divide into three types according to whether the mechanism is supposed to be (*A*) Osmotic, (*B*) Capillary, (*C*) Molecular. These types may be further subdivided:

A. Osmotic processes or flow of water due to:
    i. Swelling or imbibition ⎱
    ii. Loss of water or dehydration ⎰ of colloidal constituents.

B. Capillary changes due to:
    i. Change of interfacial tension between internal phases.
    ii. Change of electrostatic charge on a surface.

C. Rearrangement of orientated molecules:
    i. In mono-molecular layers.
    ii. In "liquid crystals."

In theories of type *A*, it is assumed that the muscle fibre contains relatively solid colloidal constituents as well as a more liquid phase and that at rest the two are in equilibrium osmotically. On stimulation the equilibrium is supposed to be upset so that the more solid phase exchanges water with the other. Either loss or gain of water might produce a shortening or a lengthening of the fibre according to the structure of the solid phase. It seems exceedingly unlikely that an exchange of water between the fibre and its surroundings is concerned. It is well known that a fatigued muscle has a higher osmotic pressure than a resting one, but this is almost certainly a secondary and minor consequence of its activity, which involves a breakdown of larger into smaller molecules.

In type *B* i, it is assumed that there are interfaces within the fibre, such as those between oil and water, at which there is an appreciable surface tension. In *B* ii, it is not necessary to assume any special type of interface, it might be something of the oil/water type or it might be simply a solid protein surface in contact with water, such as the surface of a gelatine jelly in water or the (supposed) internal surfaces of a gelatine jelly. The structure must be arranged longitudinally and be ionized, but that is all.

In type C, it is not necessary to assume anything more than some kind of determinate molecular orientation such that there are longitudinal structures in the fibre and that the architecture of the structure is labile.

The change must be in the main, if not entirely, a consequence of the production of lactic acid. That is to say, it is presumably such a change as might be produced by an increase of hydrogen ion concentration in a system consisting of protein or lipin material or both. As the protein is quantitatively the chief constituent, it is reasonable at a first attempt to look to changes in protein as the primary ones. Moreover, a considerable part of the lipin material is not ionized. The ionized substances will be *phosphatides*, of these *lecithin* has an acid dissociation constant of too large an order for its ionization to be much altered by small quantities of lactic acid; of the other known phosphatide *kephalin* it is not possible to speak with certainty, it may not be present in appreciable amount. The only known change produced in the arrangement of molecules of this type is a small change in the packing of mono-molecular films of free fatty acids at a water/air interface with large change of hydrogen ion concentration (Adam, 1922). This effect has only been observed with free fatty acids under certain conditions, not with their compounds, and it is most unlikely to be significant for our present enquiry. Attention should be paid therefore to the proteins. In the present state of knowledge the lipins must be neglected.

The reaction of most muscles is near pH 7. The isoelectric region of the proteins is on the acid side of this but not far removed from it. Consequently the proteins will be greatly affected by a small increase in hydrogen ion concentration. As the effect of bringing a protein nearer to its isoelectric point is always to cause loss of water or dehydration, alternative A i may be dismissed. Consequently the alternative A ii presents something that almost certainly happens as part of the process, i.e. some protein structure probably loses water. What is not clear is whether the process is likely to be of primary importance or of any importance. Such a process is probably responsible for the changes of *rigor mortis*, but the muscular shortening of rigor is not necessarily or even probably analogous to that following excitation of the live muscle. Meyerhof appears to lean towards a theory of type A ii. It is to be noted that a theory of this type is complementary rather than contradictory to theories of types B and C; it describes events that are likely to occur but may be subordinate in importance to events of different physical type. There is said to be histological evidence for a flow of liquid during contraction.

Hill (1925, 2, 1926, 4) has calculated, using the known data for the lactic acid liberated and tension developed, that the tension developed in a frog's muscle during a twitch is of the order of 20,000 dynes per mm.[2] of cross section; this involves either a development of surface far larger than the lactic acid liberated in a single twitch could possibly cover as a monomolecular layer, or else an impossibly large interfacial tension. So that B i may be considered as eliminated. Hill points out that a surface of un-dissolved protein ionized as an anion will carry a set of negatively charged points which will repel one another so as to tend to enlarge the surface. Neutralization by acid could discharge the surface and produce collapse. Electrostatic forces of this sort are large enough not to be numerically insignificant compared with the forces in an active muscle fibre.

Garner (1925) has suggested as a basis for the contractile process a mobile organized structure, either a film or something of the liquid crystal type, which on stimulation changes to a new type occupying a smaller area and of greater rigidity. Such a change might be brought about by very small quantities of acid if each molecule of acid was liberated at a critical point of the structure. These suggestions are in themselves not improbable but all changes of this sort are at present purely hypothetical. Nobody has ever studied a film or liquid crystal system in which a small increase of hydrogen ion concentration brings about changes such as are supposed. Hill's view has the advantage over the other possible theories that it postulates only processes of a known type which are likely to occur and to be capable of producing forces of the right order of magnitude.

We may conclude then that theories of types A i and B i are contrary to the facts. Theories of types A ii, B ii and C are not contrary to the facts as known and are therefore possible. But they are not mutually incompatible so that all or any of the changes supposed by the various theories may occur together. In our present state of ignorance as to the architecture of the muscle, Hill's theory *B* ii has the advantage that it involves no gratuitous assumptions, and the forces assumed to operate are not obviously inadequate.

We may state the case briefly as follows. On stimulation of a muscle fibre the wave of excitation passes down it; by increasing the permeability of a membrane or by some other means it causes the liberation of lactic acid from a carbohydrate source. The liberated hydrogen ions neutralize the negative charge on a surface of protein, Meyerhof's *Verkürzungsort* (or act on some other organized structure), and thereby alter the type of structure, the area of surface and the mechanical constants. This will be the fundamental change. The new structure will develop its natural length and tension slowly owing to the viscosity of the whole system. The relaxation process will consist in the restoration of the electric charge to the *Verkürzungsort* at the expense of the *Ermüdungsort* which will now in effect have neutralized the acid. In other words, contraction consists in neutralizing the electrical charge on an organized protein; relaxation in transferring the charge back to this and neutralizing the charge on unorganized protein or protein in solution. Presumably the organized protein has a higher acid dissociation constant than the unorganized.

Assuming that in any case the excitation process and the liberation of acid take place quicker than the events following, the time scale of those events will depend upon (*a*) the viscosity of the system, which will regulate the rate of development of tension, and (*b*) what may be called the "dissociation gradient," as between *Verkürzungs-* and *Ermüdungsort*, which along with (*a*) will regulate the rate of loss of tension. Muscles vary in time scale from ones which can perform about three hundred contractions per second to those which take minutes or possibly even hours for a complete contraction and relaxation, but it is conceivable that one type of mechanism such as we have supposed can accommodate even this range of variation. We have spoken so far of the single muscle twitch following a stimulation as the one type of process present. Other possibilities remain now to be discussed.

§ III. *MUSCULAR TONE*

It has been generally but somewhat lightly assumed that the phenomena of *tonus* must be based upon some fundamentally different process in the muscle. This assumption cannot at present be refuted but neither is it necessary in the present state of knowledge, as I shall endeavour to show. The subjects to be discussed naturally fall into four groups, tone in vertebrate skeletal muscle, vertebrate heart muscle, vertebrate visceral smooth muscle, and in certain types of invertebrate smooth muscle.

These will be taken in order, but first some possible misapprehensions must be mentioned. A good deal of controversy has turned upon the question as to whether muscles in tonic contraction do or do not consume more oxygen than muscles at rest. In the first place, experiments on oxygen consumption must be critically considered to decide whether the *oxygen consumed* is equivalent to the *oxygen required* for complete oxidative recovery; it is only the latter that is a measure of the *energy requirements* which are what we really want to know. If the oxygen supply is below the oxygen requirements, the experiment will probably tell us nothing. Many of the published experiments will not stand criticism from this point of view.

Secondly, assuming we know the oxygen requirements and hence the energy requirements of the muscle in tone, the next question is whether these are equivalent to or less than those of the muscle in tetanic contraction (fused twitches). The work of Hill and his school has shown (*a*) that the energy requirement under all conditions is a function of tension developed, (*b*) the energy needed to maintain a given tension in tetanus is proportional to the number of twitches per unit time needed to maintain a steady state of tension, that is, it varies inversely as the time of relaxation. In tetanus therefore *energy needed* varies as the $\dfrac{\text{tension maintained}}{\text{time of relaxation}}$. Unless the tension and the relaxation time are measured the measurement of oxygen consumption is futile.

Thirdly, it is necessary to eliminate from consideration certain states that may be loosely spoken of as tonus. It is obvious that any condition of *contracture* or *rigor* which is irreversible is not to be considered nor any change which though reversible at first insensibly passes into an irreversible effect and puts an end to the normal functions of the tissue. Thus the systolic arrest of the perfused frog's heart by excess of calcium salts is not a state of tonus, if it results in irreversible loss of function. Further there are transient states of contracture brought about in skeletal muscle by certain drugs which must be considered separately. Thus *veratrine, acetyl-choline, nicotine,* and excess of potassium salts produce a prolonged contraction of skeletal muscle. In the case of veratrine, Hartree and Hill (1922) found that there was a prolonged output of heat corresponding in time and quantity to the tension developed. The prolonged contraction with veratrine is therefore quite analogous to tetanus. The difference is that normally a stimulus opens the valve, as it were, to release the store of energy and the valve then shuts automatically, to be opened again only after a lapse of time by a fresh stimulus. Consequently repeated stimulation is necessary to maintain tension. After veratrine the stimulus opens the valve as before but the valve sticks, or if it closes opens again for a considerable time of its own accord. According to Riesser and Richter (1925) the tensions produced by acetyl-choline are small, 10–15 per cent. of the tension for tetanus. Meyerhof (1924) also says that there is very little tension and no measurable lactic acid formation. The electric change (Reisser and Steinhausen, 1922) is not oscillatory but steady and dies away as the mechanical response does. The electric change in a prolonged steady veratrine contraction appears to be similar (Hoffmann, 1912). Gasser and Dale (1926) find that in mammalian muscle the ratio lactic acid/tension is the same in acetyl-choline contracture as in tetanus, so that there can be little doubt that its effect is similar to but smaller than that of veratrine. As to the nicotine effect and that produced by potassium salts there is no evidence. They may possibly be due to some special process but it is unlikely. At any rate they cannot yet be profitably discussed.

TONE IN VERTEBRATE SKELETAL MUSCLE

As has been mentioned before, the terrestrial Vertebrate, and presumably the terrestrial Arthropod also, has to maintain a definite bodily posture in the gravitational field and that posture must be plastic in order to make movement possible. Posture is maintained by means of prolonged small tensions in the skeletal muscles which are maintained and modified reflexly. Our knowledge of the postural reflexes and of the plastic tonus of the skeletal muscle is largely the work of Magnus and his school on the one hand and Sherrington and his school on the other. The question of reflex tone has been well reviewed by Cobb (1925) (see also Fulton, 1926, Chapters XV and XVI). In the present place only certain special points will be dealt with.

It has been shown that the maintenance of tone in skeletal muscle, at any rate of some forms of tone, is associated with a definite increase of muscle metabolism (10–25 per cent. in *decerebrate rigidity*) over and above that of the resting muscles (Dusser de Barrenne and Burger, 1924). It is not, however, a tetanic contraction of the ordinary type. The energy consumption is almost certainly too small. Liddell and Sherrington (1924, 1925) find that the plastic tone of skeletal muscle is essentially of the nature of a *stretch reflex* (*myotatic reflex*). Passive stretch stimulates the *muscle spindles* and produces a small reflex stimulation of the muscle; a stimulation, that is, of a few fibres, tending to maintain the length of the muscle constant. Fulton and Liddell (1925), following a suggestion of Forbes (1922), explain the small and variable character of the electrical change found in conditions of tonic contraction or in decerebrate rigidity, which is the extreme case of reflex tone, by the assumption that the muscle fibres are stimulated asynchronously. Langelaan (1925) has made a similar suggestion. This account of the reflex tone of skeletal muscle is of such importance that it is worth considering in more detail.

It is well to bear in mind that in the vertebrate skeletal system the functional unit is not the muscle fibre or syncytium of muscle cells, as it is in visceral muscle, but the system composed of the muscle sense organs, the nerves of the reflex arc and the muscle fibres. The isolated muscle represents merely the *effector* half of the system. As the *affector* part is out of action the muscle has no spontaneous activity, and stimulation is entirely indiscriminate; it is possible to have all muscle fibres at rest or all active but no selection or control is exercised over intermediate stages. It is only by studying the intact reflex system that the whole mechanism can be elucidated. When a reflex *movement* is produced by stimulating a sensory surface the chief characteristics are (1) rapid fatigue, (2) considerable after-discharge and (3) no precise localization of response in one muscle or group of muscles. In addition (4) the electrical change observed and other effects are just such as would be expected from a rhythmic succession of impulses in the motor nerve. There is no essential difference from artificial tetanic stimulation of appropriate frequency. The characteristics of reflex tone and of the myotatic reflex are different: (1) Fatigue does not readily appear, (2) the response to stretch is practically dead beat, (3) the response is precisely localized in the group of muscle fibres surrounding the affector organs stimulated. Finally (4) the conditions are quite different from those of artificial tetanus. In all probability the energy consumption is smaller and, as mentioned already, the electrical change is small and variable, because the recording instrument is affected only by a simultaneous change in a large number of fibres.

Forbes, Whitaker and Fulton (1927), who have examined the mechanical and electrical changes, find no qualitative difference between tone in the extensor muscles of a mammalian limb and an extensor reflex in the same muscles. Even in the most vigorous reflex contraction only a few fibres appear to be active judging by comparison with the much greater effect of direct maximal stimulation of the motor nerves.

As Fulton and Liddell (1925) say: "Presumably the smallest locus or reflex unit is the group of motor fibres controlled by one 'stretch' afferent organ. At any given moment in a rigid muscle only a small number of these afferent-efferent systems are active. One would suppose that when such an afferent ending is stretched the motor fibres controlled by it would contract. If this contraction relieved the stretch on their own afferent ending thereby removing their own stimulus, other stretch afferents would be stimulated as a result of this cessation of support in the first afferent-efferent system. Or if the first suffered depression, as by autogenous inhibition arising from other types of afferents in the muscle, their neighbouring afferent-efferent systems would become active and thus maintain the state of rigidity in the muscle.

So the process might be conceived of as progressing indefinitely in endless asynchronous rotation".

This seems adequate to account for all the facts of tone in skeletal muscle without assuming any mysterious second type of "catch" mechanism or special kind of metabolism other than that observed by stimulation of the isolated muscle. The one fundamental activity of the muscle fibre is the muscle twitch. The further questions, as to whether tonic contraction is associated with some special type of fibres in the muscle and not with others, or whether innervation from the sympathetic system is concerned, may be left open for the present (see Cobb, 1925).

VERTEBRATE SMOOTH MUSCLE

Passing from vertebrate skeletal to vertebrate visceral muscle, stomach for example, we are passing from a highly specialized system to a less specialized. Functions that in the skeletal system are shared between muscle spindle, reflex arc and muscle fibres, in stomach muscle are all carried out by the muscle fibres themselves, or possibly the muscles and their nerve plexus, for it is not easy to isolate the one from the other experimentally. At any rate isolated stomach muscle (1) is spontaneously active, (2) is stimulated by passive stretch or, perhaps it is safer to say, is made more excitable. These two phenomena are also seen more unequivocally in the isolated heart, where nervous effects can be eliminated. Now it is more particularly among the visceral muscles (excluding heart) that the phenomena of tone are noticeable. The muscle of the cardiac region of the stomach, of the urinary bladder and of the walls of the arteries are outstanding cases. Among Invertebrates, the unstriated portion of the adductor muscles of the shell of bivalve Molluscs has been specially studied and it is for these cases that the idea of a special "catch" mechanism is usually invoked.

Let us first consider the case of mammalian stomach muscle about which there is a considerable body of information. Isolated strips of stomach muscle show rather different behaviour according to the region from which they are taken and according to the animal (Brown and McSwiney, 1926, 1, 2). But speaking generally the strips show rhythmical contractions superimposed upon a varying degree of tone. Either rhythmic contraction or tone may be absent or both may appear together, but the preparations showing most tone show least rhythmic movement (*fundus* region) and those showing most rhythmic movement least tone (*pyloric* region). The activity can be controlled by certain drugs. *Pilocarpine* produces a general increase of activity of whatever type is dominant in the tissue used; *atropine* is generally inhibitory. The effect of *adrenaline* is peculiar; it is generally inhibitory, but in muscle that exhibits tonic contraction it increases tone when tone is absent and diminishes it when it is present already. The muscle is also affected by the ionic concentrations of its surroundings.

The phenomena as they stand could be accounted for without assuming any special "catch" mechanism for the tonic contraction, if it were assumed (1) that passive stretch is a stimulus, (2) that there is a long relative refractory period after contraction, (3) that pure tone without rhythmic movement is due to completely asynchronous contraction; rhythmic movement without tone to completely synchronized contraction; mixed effects to partially synchronized movement. Presumably synchronization is a function of the nerve net. In pure tonic contraction it is assumed that a small number of fibres distributed throughout the tissue are active at one time. Their relaxation, when fatigue has lowered their excitability, will stimulate the most excitable by passive stretch, the most excitable being those that have had the longest time to recover since previous activity. So the activity will go on in rotation keeping the whole tissue of constant length as long as excitability and rate of recovery from the refractory state remain constant. The effect of drugs is to raise or lower the threshold of excitability. The tendency of the tissue to maintain constant length with varying tension is explained of course by the assumption that passive stretch is a stimulus. The histological observations of Lingle (1910) on heart and stomach muscle of turtle, frog and *Necturus* favour the view here advocated, that only some fibres are active at any given time.

Are there any facts which contradict this theory?

The observations of Evans (1923) on mammalian smooth muscle, in which he found that increasing the tone by *histamine*, *pilocarpine* and other drugs slightly diminished the oxygen consumption, might appear *prima facie* to contradict the theory. In these experiments the muscle in tone was allowed to shorten. Now Evans and Hill (1914, also Hill, 1925, 1) found that in *isometric* contraction the heat production (and therefore ultimately the oxygen intake) of skeletal muscle increased with length up to a maximum. On the present theory, at a constant state of tone the contractions of the fibres responsible for maintaining it will be roughly isometric so that at the shorter length less heat should be produced and oxygen consumed, although the condition is one of increased excitability. According to the results of Fenn (1924) and Azuma (1924) there should be increased oxygen consumption while the state of tone is changing, but this would not be measurable by ordinary respirometric methods. At the same length of muscle there should be increased oxygen consumption with increased tone, because *ceteris paribus* metabolism varies as total tension under isometric conditions. Bayliss (1928) calculates that the observed oxygen consumptions of mammalian smooth muscle do not necessitate the assumption of a mechanism for tone, other than tetanus, if the time relations of the contractile process are considered.

VERTEBRATE HEART MUSCLE

An organ which behaved as a whole in an all or none fashion would provide much more satisfactory evidence. The isolated heart of most animals behaves in this way, and in this organ tonic effects are, to say the least, not conspicuous. Its main function is that of rhythmic contraction and its main divergence from skeletal muscle is its spontaneous activity, or rather the spontaneous activity of certain regions. However, records of the mechanical response of the heart sometimes show small variations in the diastolic base line that cannot be attributed to changes in pressure of the perfusion fluid. Systolic arrest of the heart is not to be considered in this connection, as it is a prelude to death changes, but only changes produced while a fairly normal and vigorous beat is maintained. The phenomena of tone in the auricle of the tortoise's heart must also be excluded because these are probably due to the activity of unstriated fibres which may not behave in an all or none fashion (Rosenzweig, 1903). Perhaps the best piece of evidence is that obtained by Deseo (1926), who studied the isometric response of the frog's ventricle regularly stimulated by induction shocks at the rate of 12 per minute. The ventricle was perfused with Ringer's solution of varying calcium content and a distinct though not a large increase in diastolic tone obtained with about three times the normal concentration of calcium. At the same time the actual size of the contractions was not diminished but in some cases increased. The technique used would appear to eliminate all disturbing mechanical factors, such as complicate most perfusion experiments. These experiments show clearly that heart muscle within the limits of normal conditions can have a variable resting length. While it is probable that tonus in vertebrate visceral muscle of various types is to be explained as suggested above on similar lines to the tonus of skeletal muscle, the other possibility cannot be considered as eliminated, namely that the muscle fibres have a variable resting length on which their contractions are superimposed. A change in resting length need not involve a change in metabolism. Of course the two mechanisms may operate together. It must be understood that to call a change in tone a change in resting length is not an *explanation* of tone but merely a definition of the term.

INVERTEBRATE MUSCLE

Turning to the Invertebrates, the evidence is somewhat confused. Many soft bodied animals are capable of remaining in a contracted state for hours at a time. For instance, sea anemones, leeches, holothurians, and the worm *Sipunculus* have all been studied. Some workers have found an increased metabolism in the contracted state, e.g. Cohnheim and von Uexküll (1912) with leeches, some have found none, e.g. Parker (1922) with anemones. It seems doubtful whether the actual gas exchange of the animals in the contracted state is a good index of their metabolism, they may be simply piling up an oxygen debt. An animal with a slow metabolism can probably do this for many hours; as many mud-burrowing animals undoubtedly do (see p. 54). Most attention has been given to the Lamellibranch Molluscs. Excepting a few forms such as *Solen*, each muscle of the shells is in two parts, one clear in appearance (vitreous) and in some forms (e.g. *Pecten*) consisting of striated fibres; the other milky in appearance and always unstriated. In *Pecten*, *Ostrea* and some other genera there is only one large *adductor*. The bulk of the muscle of *Pecten* is striated and is used for the flapping movements by which the animal swims, the smaller unstriated portion maintaining the tension against the elastic ligament of the hinge when the valves are closed. On cutting out the muscles from the shell the striated portion is relaxed and responds to stimulation in the ordinary way with a rapid twitch, the unstriated portion is usually contracted, probably stimulated by injury in cutting, and generally remains in that state until it dies. By removing the whole of the rest of the animal the unstriated portion can be obtained in a relaxed and excitable state. It responds to stimulation with a slow twitch. The muscles according to Pavlov's (1885) observations on *Anodonta* are supplied with both excitor and inhibitor nerve fibres. If a *Pecten* is made to close its valves upon a wooden wedge, the wedge is held tight. On pulling it out the valves do not close at once but remain motionless. If they are gently pressed together they remain at the new position (for a time at any rate); on the other hand, against an attempt to pull the valves open a considerable tension is exerted (von Uexküll, 1912). Forcing open the valves often results in rupture of the muscle. In this tonic condition considerable tensions can be maintained for hours. The phenomena in the intact animal are similar to those of reflex tone in Vertebrates except that the molluscan muscle develops much higher tensions. Apart from this quantitative difference there is nothing to suggest a fundamentally different mechanism.

As in these animals it is easy to vary the tensions isometrically, a study of the metabolism might be of value. On the theory advocated here an isometric increase in tension must be accompanied by an increased metabolism because it must be due to the activity of a larger number of fibres at any given moment, or an increase in tension of those that are active. As the tensions developed are high in these animals it might be expected, other things being equal, that the heat production and metabolism would be large. Parnas (1910) gives measurements on *Venus verrucosa* showing that the slow adductor muscle is about 1/30th by weight of the soft parts of the animal. If the tissues have all an equal resting metabolism the metabolism of the muscle would need to be a little more than doubled to produce a 3 per cent. increase in the total metabolism. In the most regular of his experiments variations of oxygen consumption of about 2 per cent. occurred during normal periods and in some experiments much larger variations. Consequently only large changes in the metabolism could have been detected. Parnas measured the oxygen consumption during normal periods when the shells were closed and during periods when weights were applied to the valves so as to increase the tension on the muscles. It is to be noted that the increase in tension will not have been proportional to the weight used unless previously the valves were only just closed so as to balance exactly the pull of the elastic ligament of the hinge. In some series of experiments both O_2 intake and CO_2 output are recorded and the respiratory quotient ($1 \cdot 7$ in one case) shows that the oxygen requirements of the animals were not being met; i.e. they were getting as much oxygen as they could, not as much as they needed. In some experiments loading the valves produced no change but in others caused a decrease in oxygen consumption slowly rising to normal on removing the load. This is very likely the result of disturbing the animals and making them close their valves tighter. Parnas gives results of one experiment on two *Pecten jacobaea* which had their valves slightly open and presumably a better oxygen supply. Loading in this case produced an increase of 1 per cent. in O_2 intake, well within the experimental error. Parnas gives no data for the amount of slow muscle in this species but it was probably of the same order as in *Venus*. The only conclusion that can be drawn from Parnas' experiments is that no *large* increase in metabolism was produced by loading the valves. He argues as does Bethe (1911) that to produce a similar tension for a similar time vertebrate skeletal muscle in tetanus would have expended an enormous amount of energy, easily measured as increased oxygen consumption. The conclusion is correct but not necessarily relevant because the slower tissue would in any case be relatively economical. What needs to be decided is whether there is a change in muscle metabolism of the correct order for that kind of muscle in tetanic contraction, and that the experiments cannot decide. From heat measurements of Hartree and Hill and lactic acid and heat measurements of

Meyerhof we may calculate as follows. Frog's sartorius muscle maintaining tension at $15°$ C. gives out 35×10^{-6} cal. per cm. length of muscle per gm. tension per sec. total oxidative heat (Hartree and Hill, 1921). In combustion of carbohydrate 1 c.c. of oxygen produces 5 calories. Therefore the muscle consumes 7×10^{-6} c.c. of oxygen per cm. per gm. per sec. Parnas (1910, p. 488) calculates that in his experiments the slow adductor of *Venus verrucosa* used 0·0056 c.c. of oxygen per hour while maintaining 3 kg. tension. He does not state the length of the muscles but from the size of the animals used the muscles must have been about 8 mm. long. We can calculate that 8 mm. length of frog's muscle would consume 60 c.c. of oxygen per hour while maintaining 3 kg. tension, more than 10,000 times as much as *Venus*. Assuming all other things to be equal (a large assumption but the only one possible) the energy required to maintain tension is proportional to the number of times the contractile mechanism has to be excited in order to maintain a steady state of contraction; that is, it is inversely as the duration of relaxation isometrically. Therefore the relaxation time of molluscan slow muscle should be over 10,000 times that of frog's sartorius, which last we can put at 0·05 sec. Actually, according to Parnas' figures, the relaxation time of molluscan slow muscle should be of the order of 9 minutes, which is long but not longer than is possible. Marceau (1909, pp. 379, 380) gives a number of tracings of the spontaneous movements of the valves of oysters (*Ostrea edulis*) with only the slow part of the adductor muscle left. The time of relaxation is from a quarter to one hour. It is very doubtful whether Parnas' figures are correct, they are probably too low; but they are used for this calculation because they were believed by their author to provide a *reductio ad absurdum* of the view that the tension maintained by molluscan slow muscle could be of the nature of tetanic contraction. Evidently they provide nothing of the kind, if we consider the difference in time scale (neglecting other unknown differences which may exist) between frog's sartorius muscle and *Venus* slow adductor. It is assumed that *Ostrea* and *Venus* are similar and that Marceau's figures show the order of magnitude of the relaxation time.

A further calculation is worth giving; one maximal stimulation of frog's muscle (single shock) at 15° will produce 300 gm. cm. of total oxidative heat per gm. (Hartree and Hill, 1921). Dividing by 4.24×10^4 to convert to calories and by 5 to convert to c.c. of oxygen we find that a single maximal twitch requires 0·0014 c.c. of oxygen per gm. The resting oxygen consumption of frog's muscle at about 15° C., choosing a low value, may be taken as 0·02 c.c. O_2 per gm. per hour; so that a stimulus once every four minutes would just about double the rate of metabolism. This rate of stimulation is probably adequate for tetanus in *Venus* slow muscle, as will be explained shortly. Let us assume that molluscan resting metabolism is of the same order as that of frog's muscle, which is roughly true, as Collip (1921) found *Mya arenaria* at 14° C. consumed between 0·014 and 0·058 c.c. O_2 per gm. per hour; and suppose that the muscle tissue has the same metabolism as the rest. Let us assume further that the heat production per stimulus is of the same order as in frog's muscle; actually it should be less because less energy is needed to maintain than to develop tension, and because the frog's muscle is longer. It follows that the slow muscle of *Venus* can probably be kept in a state of maximal tension by tetanic stimulation while not more than doubling the resting metabolism. As the slow muscle of *Venus* is only about 1/30th of the body weight it is clear that Parnas could not expect to get a measurable increase in metabolism by increasing an already existing state of tension.

The slow adductor muscle of *Pecten opercularis* has recently been studied by Bayliss, Boyland and Ritchie (1928). The behaviour of the isolated muscle resembles that of isolated frog's skeletal muscle, allowing for the difference in time scale. While *Pecten* quick muscle works at a speed of the same order as frog's muscle, the relaxation time of the slow muscle under the tension of the elastic hinge ligament is about 100 to 1000 times longer than that of frog's sartorius; 300 times may be taken as a safe figure. This means that other things being equal the slow muscle can maintain a given tension for ten hours with the same energy expenditure that the frog's muscle needs to maintain it for two minutes. As Biedermann in 1885 published tracings of the twitch of *Anodonta* adductor muscle the relevant facts have really been known for a long time, but in the absence of a detailed knowledge of the energy exchange of frog's muscle the significance of the time factor escaped notice.

Pecten opercularis is a more active animal than most Lamellibranchs and is not accustomed to keeping its valves closed for long periods as are oysters and many others, including the group to which *Venus* belongs. From Marceau's (1909) experiments it is clear that *Pecten* muscles are generally faster in their processes than those of the oyster and similar animals, at least ten times as fast. He finds also that *Pecten maximus* in moist air can keep its valves closed for two to three days, while *Ostrea edulis*, *Dosinia exoleta* and *Gryphea angulata* can do so for twenty to thirty days. It was assumed above that to keep *Venus* slow muscle permanently contracted a stimulus once every four minutes would suffice. As eight times this rate generally suffices for *Pecten opercularis*, the assumption seems reasonable.

There are several doubtful points about the behaviour of the slow muscle of *Pecten* and other Lamellibranchs, but nothing has yet been observed which is incompatible with the views here advocated.

The really striking thing about the adductor muscles of these molluscs is not their small metabolism but the enormous tensions they seem able to produce. The "tonic" contractions of other muscles involve only quite small tensions. Marceau (1909) has found tensions of the same order as and even much larger than those developed by vertebrate skeletal muscle. Some of these results are given in Table 5. The tensions the muscles can develop are not far short of their breaking tensions.

Table 5

Tensions in Mollusc Adductor Muscles
(From Marceau, 1909)

Species	Maximum tension in kg./sq. cm.	
	Quick portion	Slow portion
Anodonta cygnea	2·0	5·2
Ostrea edulis	0·5	12·0
Tapes decussatus	1·0	12·6
Venus verrucosa	0·6	35·4
Cardium edule	1·2	25·0
Mytilus edulis	6·5	11·3
Pecten maximus	—	8·5

For comparison, human muscles can develop 7–10 kg./sq. cm.

It is not claimed as proved that the closure of the valves of Lamellibranch Molluscs is maintained by tetanic contraction of the muscle, but it is claimed that there is no justification for dismissing the possibility and assuming that a different type of process is at work. The view put forward here is that there are no facts at present known to contradict the assumption that all muscles work by means of the same fundamental process, the muscle twitch; the differences in the behaviour of different muscles is attributed to differences in the time scale of the muscle and the conditions of stimulation. If there are other processes at work they have not been demonstrated, with the possible exception of the change in *diastolic tone* of the heart.

If the tonic contraction of any muscle is to be attributed to any other process besides the ordinary one, such as change in resting length, it is well to try and see clearly what assumptions are involved. The process, whatever it is, must be (1) reversible, (2) under control of the animal independently of external changes, (3) capable of liberating energy and of doing this repeatedly. It would presumably be a change in ionic equilibrium, but if it were we have no information at present what the change could be. The frog's heart, as has been mentioned, has its resting length altered by change of calcium concentration in the perfusing fluid, but large scale changes in metallic ion concentration are not likely to be an ordinary part of the physiological mechanism, so that we cannot decide whether this effect observed by Deseo (1926) is of general significance or not.

REFERENCES

ADAM (1922). *Proc. Roy. Soc.* A, **101**, 452.
ADRIAN (1914). *J. Physiol.* **47**, 460.
— (1922). *Arch. Néerland de Physiol.* **7**, 330.
AZUMA (1924). *Proc. Roy. Soc.* B, **96**, 338.
BAYLISS, L. E. (1928). *J. Physiol.* **65**, 1.
BAYLISS, L. E., BOYLAND and RITCHIE (1928). Unpublished.
BETHE (1911). *Pflüg. Arch.* **142**, 291.
BIEDERMANN (1885). *Sitz. d. Wien. Acad.* **91**, 3. Abt. 29.
BROWN and McSWINEY (1926, 1). *Q. J. Exp. Physiol.* **16**, 9.
— — (1926, 2). *J. Physiol.* **61**, 261.
COBB (1925). *Physiol. Rev.* **5**, 518.

COLLIP (1921). *J. Biol. Chem.* **49**, 297.
COHNHEIM and VON UEXKÜLL (1912). *Zeit. Physiol. Chem.* **76**, 314.
DE BARRENNE and BURGER (1924). *J. Physiol.* **59**, 17.
DESEO (1926). *Ibid.* **61**, 484.
DOI (1920). *Ibid.* **54**, 335.
EVANS (1923). *Ibid.* **58**, 22.
EVANS and HILL (1914). *Ibid.* **49**, 10.
FENN (1924). *Ibid.* **58**, 175, 373.
FORBES (1922). *Physiol. Rev.* **2**, 403.
FORBES, WHITAKER and FULTON (1927). *Am. J. Physiol.* **82**, 693.
FULTON, J. F. (1926). *Muscular Contraction.* London and Baltimore.
FULTON and LIDDELL (1925). *Proc. Roy. Soc.* B, **98**, 577.
GARNER (1925). *Ibid.* **99**, 40.
GASSER and DALE (1926). *J. Pharm. Exp. Ther.* **28**, 287.
GASSER and HILL (1924). *Proc. Roy. Soc.* B, **96**, 398.
HARTREE (1925). *J. Physiol.* **60**, 269.
HARTREE and HILL (1921). *Ibid.* **55**, 133.
— — (1922). *Ibid.* **56**, 294.
HILL (1922). *Ibid.* **56**, 19.
— (1925, 1). *Ibid.* **60**, 237.
— (1925, 2). *Proc. Roy. Soc.* B, **98**, 508.
— (1926, 1). Croonian Lecture. *Ibid.* **100**, 87.
— (1926, 2). *J. Physiol.* **62**, 156.
— (1926, 3). *Proc. Roy. Soc.* B, **100**, 108.
— (1926, 4). *Aspects of Biochemistry,* London, p. 300.
HOFFMANN (1912). *Zeit. Biol.* **58**, 55.
LANGELAAN (1925). *Verh. Konig. Akad. Wetensch. Amsterdam,* **24**, 3.
LEVIN and WYMAN (1927). *Proc. Roy. Soc.* B, **101**, 218.
LIDDELL and SHERRINGTON (1924). *Ibid.* **96**, 212.
— — (1925). *Ibid.* **97**, 267, 488.
LINGLE (1910). *Am. J. Physiol.* **26**, 361.
LUCAS (1909). *J. Physiol.* **38**, 113.
LUPTON (1923). *Ibid.* **57**, 68.
MARCEAU (1909). *Arch. Zool. exp. et gén.* S. 5, T. 2, p. 296.
MASHIMO (1924). *J. Physiol.* **59**, 37.
MEYERHOF (1924). *Klin. Wochenschr.* **3**, 392.
PARKER (1922). *Am. J. Physiol.* **59**, 466.
PARNAS (1910). *Pflüg. Arch.* **134**, 441.
PAVLOV (1885). *Ibid.* **37**, 6.
PORTER and HART (1923). *Am. J. Physiol.* **66**, 391.
PRATT and EISENBERGER (1919). *Ibid.* **49**, 1.
RIESSER and RICHTER (1924). *Pflüg. Arch.* **207**, 287.
RIESSER and STEINHAUSEN (1922). *Ibid.* **197**, 288.
ROSENZWEIG (1903). *Arch. Anat. u. Physiol. Supp.* p. 192.
VON UEXKÜLL (1912). *Zeit. Biol.* **58**, 305.
WYMAN (1926). *J. Physiol.* **61**, 337.

Chapter IV

EFFECT OF ENVIRONMENT

ACTION OF ELECTROLYTES

The functions of the muscle cell have been discussed in the previous chapters as though it were isolated and self-contained, but like any other living cell, the muscle can function only when its composition is kept within certain limits and its surroundings are suitable. From the Protozoa which are at the mercy of the liquid surrounding them, to the higher Vertebrata whose cells are immersed in a medium of carefully regulated composition, the animal as a whole shows every degree of dependence on and independence of its environment. But the cell itself in any animal is always dependent. Ringer's great discovery that the frog's heart could beat normally in a solution of certain simple salts and that its requirements were specific has been extended to practically all living cells, though different cells have different requirements.

Unfortunately, when this much has been said, there is little more that is definite to be added. The whole subject of the action of dissolved substances in the environment of living cells is in confusion. The confusion is due partly to lack of relevant physical and chemical theory, but more also to experimental difficulties. These are: (1) It is not easy to tell whether a substance does or does not penetrate into the interior of the cell in any given case, nor (2) on what part of the mechanism of the cell it acts. (3) The effect of long-continued action of a particular environment is not always the same as its immediate action; certain factors only take effect after lapse of time. (4) Some tissues or organs are from their structure and functions difficult to investigate.

Taking these points in order: (1) Most cells, like red blood corpuscles, are readily permeable to water and to a small and strangely miscellaneous collection of substances, including dissolved gases and certain organic substances, for instance *urea*. Inorganic salts for the most part do not penetrate quickly, so that in respect to them and as far as rapid changes are concerned the cell behaves as an osmometer. The salts normally present must however penetrate slowly. A muscle fibre, for instance, contains a considerable amount of potassium which as far as is known is ionized and when the muscle hypertrophies or atrophies potassium must pass in or out. The fibre contains a much higher concentration of $K^{\cdot}$ and a much lower concentration $Na^{\cdot}$ than the tissue fluids and the blood plasma. There is some evidence that during activity a muscle loses $K^{\cdot}$ to its surroundings. It seems a reasonable assumption that the cell membrane is slightly leaky to $Na^{\cdot}$ and $K^{\cdot}$ (and other ions) but is able to perform work in transferring these ions through the membrane to maintain the normal concentration difference. It is not enough to say that the membrane is sometimes permeable and sometimes impermeable; there must be an actual process of secretion at work. As cells of the kidney and the gastric mucous membrane can secrete electrolytes and other substances against osmotic and hydrostatic pressure there is no reason why other cells should not do so too.

There is no great difference of concentration of magnesium and calcium inside and outside, so that the distribution of these ions may be the result of a simple diffusion through a somewhat leaky membrane. Considering only these four metallic ions, there is no direct evidence as to how far or how quickly altering the electrolyte composition of the environment alters the internal composition of a muscle while it remains alive. This increases the difficulty of deciding in any given case whether an effect produced by a change in the environment is entirely due to action at the surface of the muscle fibre or whether there is also a change inside, and if there is a change inside whether it is the direct result of ions passing in or out or an indirect result of purely surface changes. It is not correct to speak of the penetration of a single ion because ions of one kind cannot pass away from any region in appreciable quantities unless ions of equal and opposite charge travel with them or ions of equal and similar charge travel the other way. Strictly we ought to consider the movements of the whole system of ions. This is most obvious in the case of the hydrogen ion where the character of the anions present determines its penetration into the cell.

(2) As the *excitatory process* in muscle takes its origin at the surface it might be presumed that any effect produced without penetration is essentially an effect on the development of excitation

only and not on the contractile process which presumably depends upon internal more than surface conditions; but even this is not true (Clark and Daly, 1920).

(3) Many observers have been content to see what happened during a period of a few minutes after a change in conditions was brought about, whereas they should properly wait to see whether the change in activity was permanent or varying. Of course the difficulty of keeping the functional response of a tissue constant under any conditions makes this rather a counsel of perfection. Nevertheless, Clark (1913) has found that it is possible to keep the frog's heart in normal and reasonably constant activity for many hours, though the necessary conditions are not as simple as in Ringer's original experiment.

(4) Lastly there is the difficulty of getting equilibrium established between the tissue and the environment, which is first of all the purely mechanical difficulty of bringing the fluid into contact with each cell and secondly the difficulty already mentioned of establishing constant relations between the cell and the fluid. The most extensive and valuable studies have been made on heart muscle from cold-blooded animals. These organs have two distinct advantages: (a) Owing to the spongy character of the tissue where it is thick and its thinness elsewhere, and owing to the squeezing action of the rhythmic contraction any liquid passed through the chambers of the heart is rapidly brought in'o contact with all the cells. (b) Under constant conditions the heart's activity is constant. The frog's heart, as Mines pointed out, is an accurate timekeeper. Other types of muscle lack one or other advantage. Skeletal muscle, having no spontaneous activity, can be stimulated to give a regular response and small thin muscles come slowly into equilibrium with the liquid they are suspended in. But large muscles are very difficult to deal with by perfusion through the blood vessels. If saline solutions are used, either the capillaries are contracted so that nothing happens or they are dilated and the tissue soon becomes oedematous. Skeletal muscles are relatively insensitive to change in environment short of lethal. The smooth muscles of the digestive tract and also mammalian uterus respond well to changes of environment, but display an irregular spontaneous activity upon which experimental changes

are superimposed. According to the type of activity at the moment the excitability of the tissue varies and its response to external stimuli. For instance, relaxed muscle and muscle in "tonic" contraction give precisely opposite responses to certain stimuli. There seems to be no method of controlling the spontaneous activity of this type of muscle short of using drugs, which lower its excitability to all stimuli.

ACTION OF IONS ON THE HEART

It is impossible in a short space to consider more than a selection of the voluminous and often untrustworthy literature on this subject. The information on heart muscle, which is perhaps the most reliable, has been summarized by Clark (1927, Chapter XIV). Even where the experimental findings are beyond doubt, the facts are disappointingly miscellaneous. As far as generalization or classification is possible it is attempted in what follows. If valid at all the generalizations apply probably not only to heart muscle but to any spontaneously active muscle. In describing the influence of certain changes in the perfusion fluid it is assumed of course that all other conditions are kept constant and suitable for continued regular function: two of these are worthy of mention. Though for experiments of short duration simple salt solutions suffice, to maintain activity of the frog's heart for many hours other things must be provided; a small concentration of glucose to replace carbohydrate burnt, which seems reasonable, but curiously enough some fatty material such as crude "lecithin" or soaps of certain fatty acids, must be added to replace material washed off by long perfusion with salt solutions (Clark, 1913). (1) Anions if not toxic are of minor importance. (2) No single pure salt will keep the heart in a normal state. (3) The simplest mixture that will keep any heart going is sodium and calcium chlorides which suffice for the Spider Crab, *Maia* (Hogben, 1925). These two salts are needed for all tissues. (4) Most hearts, including *Maia*'s near relative *Homarus*, need three salts. These are usually Na$^{\cdot}$, Ca$^{\cdot\cdot}$ and K$^{\cdot}$; but some hearts, such as *Pecten* (Mines, 1911; Hogben, 1925), will function with Na$^{\cdot}$, Ca$^{\cdot\cdot}$ and Mg$^{\cdot\cdot}$. (5) The molecular ratios required by the frog's heart, which is fairly typical, are Na : Ka : Ca as 100 : 2 : 1. (6) Calcium can be fairly well replaced by strontium, and potassium by rubidium, and even in some cases by caesium and ammonium salts (Wells, 1928). About half the sodium can be replaced by any harmless solute at the right osmotic pressure. Partial replacement by other ions is possible but generally speaking the action of Na˙, K˙, Ca˙˙, and also Mg˙˙ is specific. Magnesium, though normally present, is not essential except to some marine Invertebrates. (7) The concentration of each ion can be varied without disturbing function within narrow limits only. Excess or deficiency beyond these limits brings about arrest in various ways in different tissues, as does also excess of foreign salts. As Ca˙˙ and K˙ are to some extent antagonistic the absolute concentration of both can be varied considerably provided that the ratio is kept within the right limits. (8) Ca˙˙ in excess is generally excitatory and K˙ in-hibitory, though a minimum of K˙ is necessary for normal excita-tion. (9) The hydrogen ion concentration must be between pH 8–6·5 for most tissues, but some are very tolerant of acid, such as *Maia's* heart, which will stand pH 4 (Hogben, 1925). Speaking generally increase of hydrogen ion concentration depresses functional activity. (10) The Elasmobranch Fishes need about 1·8 per cent. urea. Some other marine animals do well with some of the sodium chloride replaced by urea but it is not essential to them. (11) The body fluids of marine Invertebrates and the Elasmobranchs are isotonic with sea water ($\Delta = 2\cdot3°$ C.) but differ from it in composition. Even *Limulus* blood, which most resembles sea water, contains less Mg˙˙. The other fishes, higher Vertebrates and all land and fresh water forms have more dilute body fluids. The most dilute is that of *Anodonta* ($\Delta = 0\cdot09°$ C.), the most concentrated the marine Mammals and Teleosteans ($\Delta = 0\cdot76°$ C.).

In the main these conclusions are matters of general cell phy-siology, but a spontaneously active organ will betray small changes which otherwise escape notice. If the facts have been observed on the heart it is because that organ lends itself best to these observations. The facts as to the particular requirements of different organs and the effects of excess or defect of salts seem chaotic. Mines (1911) noticed that the varying sensitivity of the hearts of different Elasmobranchs to magnesium and to trivalent metals placed them in the order of their morphological affinities and could be correlated with differences in the normal hydrogen ion concentration of the blood. But unfortunately no other correlations of this sort have come to light. It is curious too that in one single organ, the frog's heart, the optimum hydrogen ion concentration is different in different regions (Dale and Thacker, 1913).

POTASSIUM

Wells (1928) has studied the effect of excess or defect of K˙ on the heart of *Maia* and the crops of *Aplysia* and *Helix*, which are rhythmically contracting organs of smooth muscle. The effects are qualitatively similar in each case and resemble those on rhythmically contracting vertebrate tissues, if we consider not the absolute concentration of K˙ in the perfusion fluid but its deviation from normal. Different tissues differ in their sensitivity to change of K˙ concentration and, what is probably closely connected, in the time taken to respond. It would seem that underlying quantitative differences there is a general similarity in the function of K˙.

When the frog's heart has been brought to a standstill by lack of K˙, it is still excitable and responds to an electrical stimulation with a single normal beat (Libbrecht, 1920, 1), and can be excited to give a normal or nearly normal type of rhythmical response by many substances. Not only is rubidium effective in its place and caesium partially so, but in some invertebrate tissues at any rate ammonium salts also (Wells, 1928). Libbrecht has obtained temporary recovery of excitability with adrenaline (1920, 1) and more complete recovery with 1 per cent. alcohol (1920, 2). These observations, it may be mentioned in parenthesis, are incompatible with the view of Zwaardemaker and his school that the specific action of potassium is due to its radio activity. They contend that the "potassium free" heart can be started beating by any radio-active material of activity equivalent to the potassium normally present. The experiments probably show that radio-active substances can excite tissues under certain conditions, but as they can be excited by alcohol and ammonium salts the observations do not support Zwaardemaker's thesis. Further, Clark (1922) in a series of careful experiments found no strict parallelism between the effect of potassium and of thorium and uranium salts.

Zwaardemaker (1926) has recently summarized his views with references to previous papers from his laboratory.

It is well known that frog's heart with a perfusion fluid deficient in $K^{\cdot}$ takes a long time to come to a standstill. If the normal $K^{\cdot}$ concentration is restored before the beats have stopped the immediate effect is to produce a stoppage either of the whole heart or the ventricle for a short time, after which the normal beat is resumed (Libbrecht, 1921). Similar phenomena are seen in invertebrate preparations (Wells, 1928). It follows, as both these authors argue, that normally there must exist a state of equilibrium between the inside and outside of the cell. Change of $K^{\cdot}$ concentration among other things will disturb this, but to some extent the tissue can adapt itself to abnormal conditions so that a sudden change back to normal may temporarily upset the processes just as a sudden change from normal does.

Further evidence that the excitatory action of $K^{\cdot}$ is general comes from Mayer's (1906) observations on the medusid *Cassiopea*. On excision of the sense organs the normal rhythmic movement ceased, presumably by removal of the normal "reflex" stimulation. Increase of $K^{\cdot}$ started them again. The action of $K^{\cdot}$ was inhibited by $Mg^{\cdot\cdot}$.

As has been said already the cell has to maintain a concentration gradient of $Na^{\cdot}$ and $K^{\cdot}$ across its external membrane, and it seems reasonable to suppose that it can only do so when the concentrations of both ions outside are not unlike the customary ones. If we find a tissue which does not need $K^{\cdot}$ in the perfusion fluid it is presumably because the membrane is specially "tight" with respect to $K^{\cdot}$. It is safest perhaps to say of the hearts of *Maia* and *Pecten* simply that they need unusually little $K^{\cdot}$, a conclusion supported by the fact that they are not insensitive to potassium (Hogben, 1925). Removing potassium from the fluid perfusing a tissue full of potassium and not freely permeable to that ion will not immediately alter its internal composition, but it will immediately alter the boundary conditions. Mines (1913, 1) and Clark and Daly (1920) found that a change in $K^{\cdot}$ always altered the electrical response, if it produced any effect at all. Taking the electrical response as mainly an index of excitation and conduction, and excitation as a surface process it seems likely that the primary effect of K˙ is at the surface, though it may act also on the internal processes. Generally speaking, deficiency of K˙ depresses and excess temporarily increases but finally depresses excitability.

Where, as in most experiments on the perfused heart, the organ maintains its spontaneous activity, the frequency of the beat will depend upon the ability of the tissue to excite itself and may be considered as due to two factors. These factors, which may be intimately connected but must be treated *prima facie* as separate, are the duration of the *refractory period* and the time taken for the internal excitation process to reach its threshold value. Any change in frequency may be attributed to the *excitatory process*, using the term in a wide sense to include both the recovery of excitability and the process in the tissue by which it stimulates itself. It by no means follows that any other change, such as extent and duration of the mechanical response, is not to be attributed to the excitatory process but to something else. It is true that initially the excitatory process is *all or none* but naturally the rate and course of development vary with the state of the tissue at the time of stimulation and so affect the extent and time course of the mechanical response. In order to study successfully the other processes apart from changes in excitation it is necessary to eliminate this variable. This is probably best done by artificial stimulation, as in the experiments of Clark and Daly (1920).

CALCIUM

Calcium appears in a double rôle, (1) as something no cell can do without and as concerned with the *contractile* process, (2) as antagonist of potassium. In the frog's heart (1) is most prominent, but in *Helix* and *Pecten* (2) can be seen clearly (Hogben, 1925). The effect of change of Ca¨ concentration may be to alter both the character of the external membrane and the concentration of calcium inside. It is not necessary to decide whether there is any special mechanism for the "secretion" of calcium or whether it finds its way in and out by diffusion. Increase of Ca¨ increases the *diastolic tone* of the frog's heart (Deseo, 1926). This is undoubtedly an internal effect and it may be quite different from any other observed effect. Clowes (1918) has suggested that the physiological action of Na˙, K˙ and Ca¨ is related to the fact that oil-in-water emulsions stabilized by alkali soaps can be inverted to water-in-oil emulsions by calcium or other alkaline earth salts. The idea is suggestive but too simple because the action on oil emulsions is a general property of divalent as opposed to monovalent ions, whereas the physiological action of each metal is specific; Na is not equivalent to K˙, nor Ca˙˙ to Mg˙˙ or Ba˙˙. Moreover, different tissues are differently affected. (See however Hogben, 1925, for a discussion from a different point of view.)

HYDROGEN ION

The position of the hydrogen ion, and correspondingly of the hydroxyl ion, is peculiar. They are the most mobile ions and moreover the functional activity of the tissues is constantly producing hydrogen ions. Generally tissues will function only within definite limits of hydrogen ion concentration of the external medium, with an optimum about pH 7·4. A reaction more acid than pH 6 is fatal to the frog's heart after a time, but some hearts are less sensitive. Fluids more alkaline than pH 8 generally arrest function, but this is not always a pure hydrogen ion effect. Evans and Underhill (1923) found that it may be complicated by the precipitation of calcium as phosphate. In studying the effect of the hydrogen ion the nature of the anions present is of importance because they determine the rate at which the hydrogen ion concentration of the interior is altered by change in the perfusion fluid. This was shown in the well-known experiments of Loeb on sea urchin's eggs, in which he found that the fatty acids have an effect greatly in excess of the increase in external hydrogen ion concentration they can produce, because they penetrate readily into the interior. Hogben (1925), however, found that for a given change of pH the immediate effect of hydrochloric and butyric acids on the heart of *Homarus* was the same. This would indicate either that the effect was a surface one or, less probably, that the rate of penetration of the acids was equal. The vertebrate heart behaves rather differently.

In the auricle of the rabbit's heart Andrus (1924) showed that the effect of a change of pH from 7·8 to 7·0 was different according to whether it was regulated with phosphate or carbonate. In either case there was a slowing of the beat but with phosphate it was more marked. Taking 7·3 to 7·4 as the normal pH of the tissue fluids the reaction of the tissue is a little more acid, and its excitability appears to depend, among other things, upon this difference; the greater it is, the greater the excitability. In the presence of phosphate, which does not penetrate, a change of hydrogen ion concentration outside will alter the gradient across the membrane more than if carbonate, which does penetrate, is present; at any rate the immediate effect will be greater.

It may be mentioned in parenthesis that it is artificial to speak of actual movement of hydrogen ions (or hydroxyl ions) in these fluids where their concentration is so small. Hydrogen ion concentration is merely a convenient measure of the equilibrium of all acidic and basic ions present when we do not know what those acids and bases are. A truer picture of the process is got if we conceive of a perpetual stream of CO_2 (or bicarbonate ions) pouring out through the cell membrane and of the effect on this stream of the ionic concentrations of the external medium.

Andrus makes some further observations to confirm his views that the effect of change of pH is chiefly a change in excitability. The stimulating effect of the *sympathomimetic* drugs *adrenaline* and *tyramine* was greater at the more alkaline reactions and the inhibitory effect of the *vagomimetic* drugs *acetyl-choline* and *choline* was less, as was also the effect of vagus stimulation. In the case of isolated intestinal muscle of the rabbit, which is stimulated by *acetyl-choline*, the stimulation was greater at the more alkaline reaction. Evans and Underhill (1923), working with the muscle of mammalian intestine and uterus, had already observed that within the viable range increase of hydrogen ion concentration depresses function. McSwiney and Newton (1927) found the same thing with mammalian stomach muscle. The portions that exhibit "tonic" contraction were less contracted, those that exhibit rhythmic contraction without "tone" had a slower rhythm and smaller contractions without change of base line. That all these effects are primarily changes in excitability seems the most likely explanation.

INTERNAL EFFECTS OF IONS

As has been mentioned, in studies on spontaneously active tissues the changes observed will be mainly changes in excitability; evidence as to other changes will be indirect. In skeletal muscle sufficiently strong stimuli will render changes in excitability of small consequence and enable the other changes to be discovered more easily. Clark and Daly (1920) have perfused the heart of the frog, in which the normal stimulus was abolished by a *stannius ligature*, and have stimulated electrically at a constant rhythm. In this way the effect of changes of excitability were diminished and the results made comparable to those on skeletal muscle. They confirmed Mines' (1913, 2) observation that if the frequency is altered other factors are altered too. Keeping frequency constant, however, their results were not so very different from the older observations where it was not controlled. $K^{\cdot}$ was found to produce a greater effect on the conduction of the excited state than any other change, and the effect of $Ca^{\cdot\cdot}$ to be mainly on the mechanical response. Increase of hydrogen ion concentration above the normal produced mainly a lessened mechanical response. These various effects are presumably due to internal changes not merely to surface effects. Deseo (1926) recorded the isometric response of the frog's heart, which is a much better index of the mechanical response than the isotonic contraction usually recorded. He found that an increase in the $Ca^{\cdot\cdot}$ above the normal produced an increase in diastolic tone as well as a slight increase in tension developed.

Sereni (1925) found that frog's skeletal muscle produced with decrease of $Ca^{\cdot\cdot}$ and increase of $K^{\cdot}$ a lower tension and often a residual contracture; usually a decrease in heat production but always an increase of the ratio H/T. Decrease of $K^{\cdot}$ produced a small irreversible decrease of tension and heat production with a small increase in the H/T ratio. Increase of $Ca^{\cdot\cdot}$ produced a more marked irreversible decrease of tension and heat and first of all a decrease of H/T ratio, finally an increase. It follows that both $K^{\cdot}$ and $Ca^{\cdot\cdot}$ act upon the contractile as well as the excitatory mechanism and are more or less antagonistic here also. Hartree and Hill (1924), using frog's skeletal muscle, found that increasing the hydrogen ion concentration above the normal by means of CO_2 slowed down the heat production in oxidative recovery. As the CO_2 penetrates the muscle the tissue hydrogen ion concentration will be raised *pari passu*. The change produced was of the same order as the slowing of other biological oxidation processes such as the autoxidation of glutathione (Dixon and Tunnicliffe, 1922). On the alkaline side of pH 7, however, change of reaction produces no change of velocity.

EXCITATION AND INHIBITION

The control of the spontaneously active visceral muscle requires two processes, one of *excitation* or *augmentation* and one of *inhibition*, whereas skeletal muscle with no spontaneous activity needs only one process for control, excitation. The reflex inhibition of skeletal muscle is simply the cessation of ordinary stimulation. There is an additional difference between the two types: visceral muscle is subject not only to nervous control but also to what may be called *humoral* control by means of chemical changes in the fluid in contact with it. As examples we may take the action of *adrenaline* on visceral muscle in general and the sensitivity of the contractile blood vessels and many other visceral muscles to changes of hydrogen ion concentration (Bayliss, 1923). The diminution of tone, or as I should prefer to say the depression of excitability, by increase of hydrogen ion concentrations is clearly the chief means of local control of blood flow. The functions of skeletal muscle call for delicate control by the central nervous system independent as far as possible of local changes. In consequence they would not be expected to be under *humoral* control; an expectation apparently realized in fact. Adrenaline has no action and only drastic and probably injurious changes of hydrogen ion concentration produce any marked effect. The so-called stimulation of skeletal muscle by *guanidine*, *nicotine* and *barium* salts looks suspiciously like stimulation by injury. On the other hand, all the muscles supplied by nerve fibres of the *autonomic system*, the visceral muscles that is, are affected by certain classes of drugs which are called *sympathomimetic* and *parasympathomimetic* (or vagomimetic) according as their action on the tissues is similar to excitation through one or other system of nerves. The

terms *sympatho-* and *parasympathomimetic* have been loosely used
and the different substances called by these names have different
actions. Thus, adrenaline is the most clearly and definitely sympathomimetic (considering only its action on muscle), and differs
in its effects from *ergotoxine* which acts primarily on smooth muscle
and not on heart muscle. Even among smooth muscles *ergotoxine*
is to some extent selective in its action (Barger and Dale, 1907).
Again, *pilocarpine, choline, muscarine* and *physostigmine* are reckoned
as parasympathomimetic, but they are not similar in all respects.
Pilocarpine stimulates the *uterus* of some animals though there is
no evidence for parasympathetic motor nerves to the organ. The
action of these drugs on the hearts of Invertebrates is on the whole
similar to their action on the hearts of the higher Vertebrates,
though the Invertebrates cannot be said to have anything corresponding to the autonomic system of nerves (Clark, 1927,
Chapter VIII). Thus, adrenaline accelerates and augments the
heart beat of *Maia, Homarus, Limulus, Pecten, Aplysia* and also
the contractile vessels of leeches. Muscarine, on the other hand,
depresses the activity of most invertebrate hearts (*Salpa, Helix,
Aplysia, Limulus*, but not *Daphnia*.) In some *atropine* antagonizes
muscarine action (*Helix*), in others it does not (*Aplysia*). Lastly,
adrenaline has, after a preliminary inhibition, an accelerator and
augmentor effect on the dogfish's heart, where no action of sympathetic nerves can be traced (Macdonald, 1925).

There is a connection between nervous and humoral effects.
Howell and Duke (1908) found that on vagus stimulation of the
mammalian heart there was an increase of the potassium salt in
the perfusion fluid. They suggested that the vagus-like effect of
excess of potassium on the heart was closely similar to the actual
result of stimulating the vagus. This interesting but rather
mysterious observation has not been followed up (see also Howell,
1925). However, Loewi (1921) has found that repeated stimulation
of the vago-sympathetic nerve of Amphibians and Reptiles
liberates something into the perfusion fluid which is excitatory or
inhibitory on another heart according as the original stimulation
produced excitation or inhibition. Potassium is not concerned or
is not the chief factor because the effect on the second heart is
stopped by atropine, whereas the action of potassium is not. The

meaning of these last observations will remain obscure until new evidence is forthcoming.

For the rest the most reasonable interpretation of the facts seems to be as follows. The vertebrate muscles classed as visceral are mostly self-exciting, as are also invertebrate muscles of corresponding function. In the mammalian and avian heart, which as a whole is essentially a *syncytium*, the self-exciting function has become specialized in one small region of the organ which excites all the rest. If we leave hearts out of account or consider a heart as a single muscular unit, it is probably safe to say that all visceral muscle is self-exciting. In saying this there is no need to decide whether the self excitation proceeds from the muscle fibre itself or a nerve ending, or even a nerve net where such exists. The important point is that as an organ in the body or as a structure that can be studied experimentally outside the body it is spontaneously active in the absence of connections with the central nervous system. If there is a self-exciting mechanism its excitability can be either increased or diminished. According as the activities of individual units in the tissue are synchronized or not, the increased excitability will appear as accelerated rhythm and greater amplitude of beat (heart) or as increased "tone" (blood vessels and some parts of alimentary canal). Similarly diminished excitability appears as slowing of rhythm and smaller amplitude of beat or as diminished tone. These changes may be produced (a) by nervous stimulation, or they may be produced by two different types of humoral control, (b) by electrolytes, (c) by certain organic substances of very great and specific physiological activity, or (d) the internal state of the tissue such as changes due to stretching or to fatigue or rest. Of course all four factors may operate together, and any of them may be either excitatory or inhibitory. Among the higher animals most but not all of these muscles have a double nerve supply (a) excitatory and inhibitory. But apparently there may be a single supply, either excitatory or inhibitory. Thus many arteries seem to have only *constrictor* or excitatory nerves, while the hearts of fishes seem to have only an inhibitory nerve supply. The action of factor (b) seems to be general and not specific. All types of visceral muscle are depressed by increase of hydrogen ion concentration and excited by diminution, as long as the concentrations are within certain limits. The action of hydrogen ions may be called "natural" or "normal" humoral control because changes that are found experimentally to produce appreciable effects are liable to occur in the living body and to be of importance under ordinary conditions. It is unlikely that in the normal intact animal there are sufficient variations in potassium or calcium ion concentrations or in other ions to make any difference, so that the effects produced experimentally by changes of $K^{.}$ or $Ca^{..}$ may be called "artificial" humoral control, though these ions probably act upon the same part of the mechanism as the hydrogen ion.

The action of drugs (c) differs from that of electrolytes in being more specific, so that different muscles are differently affected by the same drug. The fact that in many cases the differentiation corresponds roughly to the action of the two parts of the autonomic system justifies to some extent the use of the terms *sympatho-* and *parasympathomimetic*. If stimulation of the autonomic nerves results in the liberation of active substances (Loewi, 1921) the fact that drugs can imitate the process will depend upon their chemical resemblance to these active substances and how far they fit into the chemical structure of the excitatory mechanism of any special tissue. From the fact that adrenaline depresses some tissues and excites others while increase of hydrogen ion concentration always depresses we may perhaps conclude that the drug and the ion act on different parts of the mechanism. Among the drugs we may distinguish "normal" ones such as adrenaline and probably *histamine*, which are present in the intact body, and the other "artificial" ones which are plant alkaloids.

REFERENCES

ANDRUS (1924). *J. Physiol.* **59**, 361.
BARGER and DALE, H. H. (1907). *Biochem. J.* **2**, 240.
BAYLISS, W. M. (1923). *The Vaso-Motor System.* London.
CLARK, A. J. (1913). *J. Physiol.* **47**, 66.
— (1922). *J. Pharm. Exp. Ther.* **18**, 423.
— (1927). *Comparative Physiology of the Heart.* Cambridge.
CLARK and DALY (1920). *J. Physiol.* **54**, 367.
CLOWES (1918). *Proc. Soc. Exp. Biol. and Med.* **15**, 108.
DALE, D. and THACKER (1913). *J. Physiol.* **47**, 493.
DESEO (1926). *Ibid.* **61**, 484.

DIXON and TUNNICLIFFE (1922). *Proc. Roy. Soc.* B, **94**, 266.
EVANS and UNDERHILL (1923). *J. Physiol.* **58**, 1.
HARTREE and HILL (1924). *Ibid.* **58**, 470.
HOGBEN (1925). *Quart. J. Exp. Physiol.* **15**, 263.
HOWELL (1925). *Physiol. Rev.* **5**, 161.
HOWELL and DUKE (1908). *Am. J. Physiol.* **21**, 51.
LIBBRECHT (1920, 1). *Arch. Int. Physiol.* **15**, 352.
— (1920, 2). *Ibid.* **15**, 446.
— (1921). *Ibid.* **16**, 448.
LOEWI (1921). *Pflüg. Arch.* **189**, 239; **193**, 201.
MACDONALD, A. D. (1925). *Quart. J. Exp. Physiol.* **15**, 69.
MAYER (1906). *Publ. Carnegie Inst. Wash.*, No. 47.
MCSWINEY and NEWTON (1927). *J. Physiol.* **63**, 51; **64**, 144.
MINES (1911). *Ibid.* **43**, 467.
— (1913, 1). *Ibid.* **46**, 188.
— (1913, 2). *Ibid.* **46**, 349.
SERENI (1925). *Ibid.* **60**, 1.
WELLS, G. P. (1928). *Brit. J. Exp. Biol.* **5**, 258.
ZWAARDEMAKER (1926). *Ergeb. d. Physiol.* **25**, 535.

APPENDIX

The most recent work by Hill (1928, *Proc. Roy. Soc.* B, **103**, 117, 138, 163, 171, 183), using improved technique, has cleared up some doubtful points in our knowledge of the processes in active frog's muscle and provided further evidence for the correctness of the view that lactic acid is the one essential factor in the contractile process.

The *anaerobic delayed heat* is not part of either the contractile or the oxidative recovery process. It is apparently a prolonged increase in the resting heat production which can be obtained by stimulating under conditions of oxygen want. It is not found in a series of single twitches but only in tetanus.

The mean value for the ratio Tension-Length/Heat under anaerobic conditions is 6·16 by direct measurements, agreeing with Meyerhof's value 6·14 obtained indirectly by way of lactic acid estimations. The *total oxidative heat* is 2·07 times the *initial* or *anaerobic heat*. From this figure the ratio of lactic acid oxidized to lactic acid liberated is calculated as 1/4·81, in good agreement with 1/4·7 obtained by Meyerhof and Schulz by the most direct method (cf. pp. 30 and 36 of text).

INDEX

Printed in the United States
By Bookmasters